Just Transitions

Series Editor

Raphael J. Heffron, University of Pau and Pays de l'Adour, Pau, France

Alicia Phillips • Laura Kaschny
Editors

Global Perspectives on the Circular Economy in the Just Transition

palgrave
macmillan

Editors
Alicia Phillips
Université de Pau et des Pays de l'Adour
Pau, France

Laura Kaschny
Maastricht University
Maastricht, The Netherlands

ISSN 2731-6041 ISSN 2731-605X (electronic)
Just Transitions
ISBN 978-3-032-18341-5 ISBN 978-3-032-18342-2 (eBook)
https://doi.org/10.1007/978-3-032-18342-2

This Palgrave Macmillan imprint is published by the registered company Springer Nature Switzerland AG
The registered company address is: Gewerbestrasse 11, 6330 Cham, Switzerland

Forewords

This book is a real success, given the scale of the challenge. In a workshop co-organised by the United Nations University, Institute on Comparative Regional Integration Studies (UNU-CRIS), Erasmus University Rotterdam and Université de Pau et des Pays de l'Adour, the aim was to bring together for two days people who, although they shared a common interest (academic and undoubtedly civic) in energy justice, did not know each other before the workshop. These individuals had to work on topics they had barely chosen and produce a chapter together. The idea was to show that it was possible, through collective intelligence, to write high-quality chapters. And that it is possible, by offering coordinated work sessions, to produce a coherent work.

The finished product, this book, proves that it was not only possible but also relevant. The chapters drawn from these reflections, the introduction and the conclusion, show that environmental justice is an issue and a requirement. This requirement can be specified and qualified; we can be demanding in our expectations and qualify how these expectations can be met. This book teaches us that we can be ambitious in terms of collective intelligence and demands for justice in a climate-challenged and unequal world.

Xavier Arnauld de Sartre
Directeur de recherche au Centre National de la recherche scientifique

The International Early Career Writing & Research Skills Workshop (April 2025, Bruges), an international gathering of emerging scholars, served as an incubator for scholarly dialogue and interdisciplinary collaboration, fostering academic growth and forging lasting connections across disciplines. That transformative experience ultimately gave rise to the present volume, a vibrant interdisciplinary collection reflecting the collaborative spirit and intellectual momentum generated by the workshop. This collection showcases exceptional scholarly quality, diversity, and global scope. Chapters contributed by early career researchers from around the world address a wide array of issues—from energy justice, fast fashion, and nuclear policy to hydrogen initiatives and just mining practices—all under the unifying theme of advancing a circular economy within a just transition. The collaborative spirit and academic depth nurtured by the workshop resonate through these pages, exemplifying the value of interdisciplinary dialogue and global collaboration in shaping a more just and sustainable future.

Professor Dr. Róbert Szuchy
Vice-Rector for Education
Institute of Private Law Studies, Department of Civil Law
Faculty of Law of the Károli Gáspár
University of the Reformed Church in Hungary

Witnessing more scholarship on the just transition is always good news. As the world faces more climate extremes, more and more action is needed. Further, more innovative policy is needed and more policy risk is necessary. Countries are becoming too risk averse in policymaking but ambition is needed to achieve a sustainable world. It is in this context and very welcome that early career scholars from a variety of countries around the world have come together to produce an excellent edited collection on the just transition and the circular economy.

Indeed, the just transition and the circular economy is an important topic, and this edited collection is one of the first more practical studies on the topic. Its importance is that it is connected with all of our livelihoods and for everyone all over the world. The idea for this text germinated in September 2024, and it has been a pleasure to see the idea transform from then through to a conference at the United Nations University through to this publication. Congratulations are due to the editors, all the chapter authors and all who supported this project. It demonstrates that progress can happen with collective action in a short time. Upon reading, reflect and consider how you can contribute to the world's just transition journey.

Professor Dr. Raphael J Heffron, Abu Dhabi
Professor and Dean of the College of Law,
Abu Dhabi University

Preface

The research journey for this book began in September 2024. Thereafter it bloomed from a culmination of assigned topics from the International Early Career: Research & Writing Skills Workshop held at the United Nations University, Institute on Comparative Integration Studies (UNU-CRIS), Bruges, Belgium, from April 9–11, 2025. Convened by early career researchers Alicia Phillips, Laura Kaschny and Kevin Le Merle, the workshop was designed to equip early career researchers with the skills and guidance necessary to conduct sound research with the aim to publish and to promote a strong network of collaboration between early career researchers interested in interdisciplinary research, particularly fostering exchange on the topics of sustainability, energy justice, climate justice and a just transition. In doing so, expert scholars were invited to guide participants on writing and research, culminating in these chapters for publication, which were identified from a research analysis of the existing literature as key topics for further research in the context of the just transition to a circular economy.

Few articles exist on the topic, and are overtly theoretical when dealing with the circular economy and the just transition, primarily because they do not take the important aspect of future proofing and risk management into consideration.

As early career researchers ourselves, we are grateful to the institutions that have supported this project and to all the authors who have contributed. We hope you look forward to reading the research from across the world and interacting with issues relevant today.

7 November 2025 Alicia Phillips & Laura Kaschny

Acknowledgements

This book owes its existence to the authors and partner institutions, without whose support it would not have existed. To all the early career researchers who only met for the first time in April 2025 at the 2nd International Early Career Conference and who have produced collective chapters – we did it!

This is a prime example of the power of collective collaboration, let us continue to be sources of encouragement in this academic journey. The power of collective action on research is demonstrated here and we hope others are inspired.

The co-editors would like to express thanks to our partner institutions who supported and continue to support us in the organization of the International Energy Justice Early Career Conference series. For this, our second year being able to commune with various global scholars in Bruges, Belgium, we would like to thank:

- The Centre National de la Recherche Scientifique (CNRS), France;
- The United Nations University, Institute on Comparative Regional Integration Studies (UNU-CRIS), Belgium;
- Erasmus University Rotterdam, the Netherlands;
- Károli Gáspár University of the Reformed Church in Hungary; and
- Fundação de Apoio à Pesquisa, ao Ensino e à Cultura (FAPEC), Brazil.

To our esteemed keynotes at the workshop, who inspire and continue to inspire early-career researchers globally:

- Professor Emeritus Dirk Inze;
- Professor Jennie Stephens;
- Dr Ryan Scarrow and;
- Rachael Ballard.

Also, a special thank you to Kevin Le Merle for working with us to organise and convene the workshop at UNU-CRIS – it was a pleasure, and we look forward to further collaborations in the future!

The co-editors also wish to express thanks to everyone who believed and supported us in this project from inception to completion and who also dedicated their time to facilitate the Workshop sessions: Professors Xavier Arnauld de Sartre, Raphael Heffron, Leonie Reins, Philippe de Lombaerde, Mr Andrew Dunn and Dr Frank Mattheis.

Last, but definitely not least, to our families and partners, Raphael and Magnus, for their enduring patience and unwavering belief in our abilities.

Contents

Notes on Contributors

Alase Adetokunbo is a Sessional Instructor at the Faculty of Law, University of Manitoba. He is a PhD graduate of the School of Law, Western Sydney University, Australia. His research investigates the suitability of alternative dispute resolution to environmental dispute resolution from the mining sector perspective. Prior to his PhD, Adetokunbo was a tenured lecturer with Lead City University, Ibadan, Nigeria.

Syed Sajad Ali Shah is an MSc student in Geography of Environmental Risks and Human Security at the UNU-EHS, University of Bonn. He holds a bachelor's degree in Disaster Management from the University of Peshawar. He worked for three years with mountainous communities in northern Pakistan on disaster response and climate change adaptation, preparedness, and green energy initiatives. His interests include community resilience, sustainable development, and integrating renewable energy into climate adaptation and disaster risk reduction.

Beatriz Arnal-Calvo is a doctoral candidate at the University of Brighton and a member of the Centre for Applied Philosophy, Politics and Ethics (CAPPE). Her research explores the intersections between climate, energy, gender and peace. Specifically, she focuses on the ecological, gender and security dimensions of fossil-fuelled militarism. She is an intersectional feminist, environmentalist and pacifist, and an active member of the Women's International League for Peace and Freedom (WILPF). Previously, she worked as a humanitarian practitioner in settings of armed conflict and forced displacement in Kenya, Uganda, Palestine, Iraq, Morocco, Tunisia and the Sahrawi refugee camps in southwest Algeria.

Emiliano Castillo Jara is a PhD researcher in the Department of Human Geography and Environment Sciences at the University of Trier, Germany. Combining research on energy geographies, energy justice, and political ecology, his work focuses on conflicts over fossil fuel infrastructures in North America.

Sonia Chabane (she/her) is a PhD candidate within the REPI and an assistant within Université Libre de Bruxelles (ULB)'s Political Science Department. This PhD project examines how expert networks organise the humanitarian sector's ecological transformation, especially in the Mediterranean region. Prior to joining ULB, Sonia managed the supply of external academic expertise to all the European Parliament's Committees in the EU's foreign affairs at the Trans-European Policy Studies Association (TEPSA). She holds three master's degrees in political science, human rights, and international relations from Saint Joseph University of Beirut (2021, valedictorian) and Sciences Po Toulouse and University-Toulouse-1 Capitole (2022), focusing on UN humanitarian programmes, the financialisation of aid, and frontline community organising in Turkey, Jordan, Syria, and Lebanon. Since 2023, Sonia has been a Global Fellow at Brown University's Centre for Human Rights and Humanitarian Studies and actively joined JUSTES, an FNRS interdisciplinary research group in Belgium, which focuses on establishing bridges between social and ecological justice.

Silvia Ciacchi is a PhD candidate at Erasmus University Rotterdam, researching sustainability due diligence from a legal and economic perspective. She has a background in comparative European and international legal studies, economic law and business and human rights. Her research interests include the process of institutionalisation of corporate social responsibility in European legal systems and the application of interorganisational theory on value chains for the assessment of the legal design of GVC sustainability regulation.

Suddha Srimoyee Das holds a Postgraduate degree in Public Policy and Human Development with a specialisation in Risk and Vulnerability from the United Nations University - Maastricht Economic and Social Research Institute on Innovation and Technology (UNU-MERIT) and Maastricht University. Her background is in Sociology, Global Studies, Governance and Development Studies. Her research interests lie in disaster risk reduction, social justice, and intersectionality and holistic development.

Filiz Doğan is a political scientist and PhD graduate in EU security from Leiden University. Her research focuses on European foreign and security policy, with particular interest in the EU's strategic autonomy and broader security studies. She currently serves as a committee member of UACES and has an extensive record of academic service, including active involvement in the UACES Graduate Forum. Alongside her research, she has taught political research methods and EU external relations.

Ari Rogerio Ferra Júnior holds a PhD in Civil Law from the University of São Paulo and the University of Camerino (dual degree). His research interests include energy law, private law, sustainability, energy justice, and energy communities.

Carmen Freed Huici is a Chilean PhD student at The Bartlett Development Planning, University College London (UCL). Her research focuses on the relationship between social housing, energy vulnerability, and gender experiences. She currently works as a research assistant on the project "Exploring the Extraction–Capital–Transition Nexus in the Antofagasta Region" (FONDECYT 1251118). She is an architect and a master's in Sustainable Architecture and Energy from the Pontificia Universidad Católica de Chile (2019); her academic work concentrates on addressing issues of urban housing, energy poverty, sustainability, and gender.

Tareq Yacoub Helou is a PhD candidate pursuing a double degree in International Relations at the University of São Paulo and King's College London. He holds a Bachelor of Science in Economics from the Federal University of Paraná and a Master of Science in International Administration and Global Governance from the University of Gothenburg. His research focuses on the political economy of multilateral development finance, regional integration, the BRICS, the United Nations Security Council, and EU–Latin America relations.

Gianna Giardini is a doctoral researcher at IUSS Pavia and at the University of Camerino, Italy. Her research explores the intersection of energy, sustainability and private law, with a particular focus on energy communities. She is the author of several publications on electricity market design and has been a speaker at national and international conferences. She was a visiting researcher at the European Commission's DG ENER – Unit for Renewables and Energy System Integration Policies – where she supported the legal team responsible for overseeing the transposition of the Renewable Energy Directive across EU Member States. Alongside her academic work, she also practises as an energy lawyer.

Nils Grolimund is a doctoral researcher at the University of Lausanne – Institute of Geography and Sustainability (IGD), working on justice and equity in the Swiss energy transition. He has a background in geography, international relations, political science, and Global Studies. His research interests include justice and policy and their connections to degrowth as an avenue towards an environmentally just future.

Tom Hambley is a social scientist with an extensive qualitative research background exploring place relationships and engagement for renewable energy technologies. He completed his doctoral thesis exploring place relationships during engagement for geothermal projects from the project developer perspective in 2024 and is currently a postdoctoral research fellow at ARU's Global Sustainability Institute, working on the cross-faculty Sustainable Futures research and innovation theme.

Paola Jimenez-Casanova is a Mexican lawyer specialising in energy and environmental law. She is currently a PhD researcher at the Universitat Rovira i Virgili (URV), where she studies the legal challenges of hydrogen network planning within the European Union. Her work forms part of the project "Training for a Hydrogen Economy Renewable Energy Society in the Anthropocene" (THERESA), funded by the EU's Horizon Marie Skłodowska-Curie Actions (MSCA). Her professional experience spans roles as Regulatory Manager, overseeing legal compliance and permitting across the company's energy businesses; in-house legal advisor; legal analyst in a consultancy; and research assistant during her law studies. Paola holds an LLM in Energy and Climate Law from the University of Groningen.

Laura Kaschny is assistant professor of EU law. Her research sits at the intersection of energy law, economic regulation and administrative law, with a particular focus on how non-economic values are articulated within frameworks of economic regulation. She researches these questions in the context of societal and just transitions, tracing how value claims travel across institutional boundaries vertically (such as between EU institutions, Member States and national regulatory authorities) and horizontally (between energy law and connected legal fields such as environmental law, competition law and administrative law). Her work highlights how these dynamics shape competences, regulatory design and institutional legitimacy.

Kaschny's research has been published in high-quality peer-reviewed journals such as Transnational Environmental Law and International & Comparative Law Quarterly, as well as with leading academic publishers including Palgrave Macmillan, Springer Nature and Cambridge University Press. Her work has contributed to discussions in EU policy frameworks and has been cited in notable scholarship in her field. She was a National Nominee for the 2023–2024 Frontier Planet Prize, and she has been acknowledged as a Future Energy Leader by the Dutch Chapter of the World Energy Council.

During her PhD at Tilburg University, entitled 'EU Energy Law and Energy Justice: A Substantive, Procedural and Institutional Perspective', Kaschny examined how energy justice engages with Dutch and EU energy law. Following her PhD, she conducted postdoctoral research at Erasmus University Rotterdam, further developing her expertise in energy law and the regulation of societal transitions. She has participated in interdisciplinary research projects and collaborative consortiums such as NEON, HOLON, and V2G-Quests, working alongside engineers, social scientists and economists. These projects investigate key aspects of the energy transition in the Netherlands and across EU Member States, including the intersection of law and technology and their societal impacts.

Kaschny teaches in the field of substantive EU law and supervises student theses across a wide range of EU law topics. By invitation, she has repeatedly delivered lectures at the Academy of European Law (ERA) and the Council of European Energy Regulators (CEER) in the field of energy and environmental law.

Sil Katrien Maslov-Lescrauwaet is a research intern at the United Nations University-CRIS and a Flanders Trainee Programme grantee, specialising in environmental governance. With Chilean-Belgian roots, she brings a unique cross-regional perspective to just transitions research. She holds a master's in Conflict and Development Studies (summa cum laude) from Ghent University, and has published on regional seaweed aquaculture policy, and Amazon disaster risk management. Her work integrates Latin American political ecology with multispecies approaches to sustainability governance.

Hazal Mengi is a PhD candidate in the Department of European Union at Istanbul University in Türkiye. She is currently conducting research as a visiting PhD scholar at the Department of Spatial Planning and Environment, University of Groningen, the Netherlands, with the support of the Jean Monnet Scholarship Programme. Her research focuses on hydrogen, socio-technical transitions, and energy justice, particularly within the context of the European Union.

Strahinja Obrenović is a teaching assistant at the University of Belgrade. He graduated from the Faculty of Political Science in Belgrade and holds a master's degree in International and European Studies, specialising in International Humanitarian Law and Human Rights Law. His PhD thesis focused on the regulatory framework of the European energy market. Throughout his academic career, he has received numerous awards and scholarships both in Serbia and internationally, including the Florence School of Regulation Annual Scholarship. In 2018, he interned with The Negotiating Team for Serbia's EU Accession Negotiations within the Government of the Republic of Serbia. Additionally, he delivered a lecture on energy security and solidarity to participants of the Regional Strategic Leadership Course organised by the University of Defense in Belgrade.

Laura Peña is a PhD researcher at the Centre for Environment, Economy and Energy at the Brussels School of Governance, Vrije Universiteit Brussel. With a background in law, economics and art, her work combines legal theory and empirical methods to study environmental governance, corporate accountability, and sustainability in global value chains, participating a.o. in the EU-funded LIFE4EPR project. Her doctoral research focuses on Extended Producer Responsibility schemes and digital product passports for textiles. Alongside her PhD, Laura is an entrepreneur and founder of PARCE art space in Ghent and works as a freelance consultant in law and economics.

Alicia Phillips is a PhD researcher at the Université de Pau et des Pays de l'Adour (UPPA), France. As part of the TRansitions Energétiques et Environnementales Laboratory (TREE) & CNRS, Alicia's research explores the nexus between human rights, climate change and energy justice across jurisdictions such as Latin America, the EU, Africa, and Asia-Pacific.

A lawyer by profession, Alicia completed her academic and professional legal studies in Barbados and Trinidad & Tobago, the West Indies. She has published in premier journals including the International Energy Law Review, Journal of Energy & Natural Resources Law, Springer Nature, Journal of Urban Regeneration & Renewal, Sustainable Earth Reviews as well as The Conversation-Europe. Currently, she is also an editorial board member for Renewable and Sustainable Energy Reviews, and a visiting lecturer at the Károli Gáspár Reformed University in Hungary. In the past she has lectured in master's level courses at the Solar Academy- Université Savoie Mont Blanc and the Université de Pau et des Pays de l'Adour in France. She has also completed consultancy work on human rights and the extractive industries, and project-managed multiple events in these areas.

Kathlen Schneider is a Brazilian researcher pursuing a PhD at the Science Policy Research Unit (SPRU), University of Sussex, with a focus on just energy transitions in Brazil. She holds master's degrees in Energy and Climate Policy (Sussex) and Civil Engineering (Federal University of Santa Catarina), with early research on energy efficiency in buildings, solar photovoltaics, energy communities, and energy access in the Brazilian Amazon. She co-founded and coordinates the Brazilian Women's Network in Solar Energy (Mesol Network), which aims to promote gender equity in the energy sector in Brazil, and also coordinates the LAC Research Network at Sussex, a network that fosters collaboration on Latin American and Caribbean studies across the university.

Francisca Soto Montverde is a PhD candidate at Dickson Poon School of Law, King's College London. Her research explores the ways in which Transnational Environmental Law addresses the socio-environmental implications of decarbonisation in developing countries by looking at the global environmental governance of the lithium industry in Argentina, Bolivia and Chile (Lithium Triangle).

Klaudia Szabelka is a doctoral researcher at Glasgow Caledonian University, affiliated with the Mary Robinson Centre for Climate Justice. She is also a researcher at the University of Glasgow's School of Law. Klaudia holds an LLM in International Law and Security and an MA (SocSci) in Politics and Economics. Her interdisciplinary work draws on legal, policy, and social science perspectives to critically examine justice implications in resource governance and emerging technologies.

Carlotta Terhorst is a research associate in the Socio-Ecological Transformation (SET) research group of the Linz Institute for Transformative Change. She completed her bachelor in Cultural Studies at the University of Maastricht, has master's degree in International Studies from the University Institute of Lisbon and recently worked as a research assistant at the Max Planck Institute for the Study of Societies (MPIfG) in Cologne. Her research focuses on climate conflicts and distributional aspects of climate policymaking.

Päivi Tikkaokoski started as a doctoral researcher in the Urban Environmental Policy Group at the University of Helsinki in 2023. Before that, Päivi has worked as a researcher, policy officer, volunteer, and civil servant, which have given me valuable practical experience and insight into the climate change science-policy interface. Her PhD (based on the regional development project VILKKU) looks into private sector agency in climate risk assessment and adaptation, and contributes to discussion of just resilience in adaptation governance. She has graduated with an M.Soc. Sc in International Relations from the University of Tampere and MSc in Carbon Management from the University of Edinburgh.

Rachel Walters is a Ph.D. candidate with the Monash Sustainable Development Institute (MSDI) and the School of Social Sciences at Monash University, Melbourne, Australia. In her doctoral studies, Rachel is exploring key questions around the interplay between place attachment and energy justice in Australian community energy initiatives. Rachel holds a Graduate Diploma in SocSc in Psychology and a BA (Hons) in Human Geography. She has published in premier International Energy Journals, including Energy Research & Social Science (ERSS) and Environmental Research: Energy.

Shakya Wickramanayake is a Sri Lankan lawyer, currently pursuing a PhD in law & regulatory governance at the Tilburg Institute of Law, Technology, and Society (TILT) in the Netherlands. Her PhD, which is part of the Dutch Research Council-funded project 'Via Augusta', explores the role public values and energy justice can play in guiding infrastructure operators through the complex EU regulatory landscape created by the digital, green and energy transitions, and how these concepts can assist in facilitating cross-sectoral collaboration in the Netherlands. She holds an LLB from the University of London and an LLM in Law & Technology from Tilburg University.

Zamira Xhaferri is a University Lecturer in EU Law and a Jean Monnet Coordinator for ENLARGEU (2022–25) & European Climate Imaginaries (2023–26). She is an editorial board member of the European Journal of Risk Regulation (Cambridge University Press), a research affiliate at the CEU Democracy Institute in Budapest and a member of the IUCN World Commission on Environmental Law. Zamira holds a PhD degree in European Law from Maastricht University, an advanced LL.M. degree in International and European Economic Law from Maastricht University and a double LL.M. degree in Law and Economics from Erasmus University Rotterdam and Hamburg University.

Clelia Zardini is a PhD candidate at the Università Statale di Milan, Italy. With a background in international relations and social policies, her primary research interest lies in bottom-up social innovation. Since late 2023, she has been focusing her PhD research on the challenges faced by Renewable Energy Communities (RECs) in contributing to energy democracy.

List of Figures

List of Tables

PART I

Introduction

CHAPTER 1

Incentivising Circular Economy Strategies Towards a Just Transition

Alicia Phillips and Laura Kaschny

Abstract As a defining goal of our time, the just transition has soared in popularity from its first mention in the 1970s regarding workers' rights to its inclusion in climate change discussions and policies today. While numerous definitions exist for the concept of a just transition, central elements commonly include inclusivity, safeguarding workers' rights and environmental sustainability and 'not leaving anyone behind' during the shift to low-carbon economies. Central to the circular economy is the concept of a just transition, achieved through sustainable governance that focuses primarily on the elimination of waste. Naturally, the connection between the two strategies is obvious, as both are grounded in sustainability, with the just transition focusing on equity, fairness and inclusion and the circular economy on resource efficiency, regeneration and the elimination of waste. As governments initiate policies to propel the just transition, circular economy strategies work in tandem by promoting social equity, both always with justice at their core. While the circular economy provides

A. Phillips
UMR TREE, CNRS, Université de Pau et des Pays de l'Adour, France
e-mail: alicia.phillips@univ-pau.fr

L. Kaschny
Maastricht University, The Netherlands
e-mail: laura.kaschny@maastrichtuniversity.nl

A. Phillips, L. Kaschny (eds.), *Global Perspectives on the Circular Economy in the Just Transition*, Just Transitions,
https://doi.org/10.1007/978-3-032-18342-2_1

the practical mechanisms (the how), the just transition provides the ethical framework to make sure that sustainability is achieved for all involved, not only for current but also future generations. The focus of this chapter will be to provide the aims of the book along with how they integrate with the corresponding chapters and to present the preliminary study of circular economy strategies of the '3Rs', that is, *reform*, *reuse* and *reduce*, as targeted approaches towards the just transition.

Keywords Just Transition • Circular Economy • Sustainability • Equity • Inclusion

1.1 Introduction: Redefining the Pathway to the Just Transition Through the Circular Economy

The purpose of this book is to advance and highlight the importance of the need for the 3R's of the circular economy – 'Reform', 'Reuse' and 'Reduce' – within a just transition framework. These circular economy strategies are targeted to enhance the just transition by:

- *Reform*: As the backbone of the operation of society, reform looks at the ways just transition issues can be seen via the lens of the circular economy that is restructuring, reorganising and transforming societies;
- *Reuse*: Focuses on how products are being redesigned by maximising resource efficiency and sustainability[1] and includes recycling and refurbishment but also managing how we consume products and;
- *Reduce*: Offers a range of strategies to lower greenhouse gas (GHG) emissions in the fight against climate change, such as reducing waste and pollution via, for instance, employing renewable energy initiatives.

As a reminder, it is essential to recognise that the text's primary objective is to advocate for the adoption of circular economy strategies to facilitate a just transition. In this context, it is necessary to appreciate that the circular economy is not just a waste management strategy; it is a systemic

[1] Ahmed, W. *et al.* 2024. Circular economy: Extending end-of-life strategies. *Sustainable Production and Consumption*, *51*: 67–78. https://doi.org/10.1016/j.spc.2024.09.003

rethinking of how we design, produce and consume.[2] However, many definitions and principles exist, and also the concept is interpreted and implemented in a variety of ways.[3] Nevertheless, there lies great value in a focus on promoting value maintenance and sustainable development; the creation of environmental quality and economic development; and overall, increasing the social equity to the benefit of current and future generations; and this is all enabled by an alliance of stakeholders (industry, consumers, policymakers and academia) and their technological innovations and capabilities.[4] Altogether, advancing just energy transition goals requires close collaboration across a complex ecosystem of actors to adopt a human-centred approach so that no one is left behind.[5]

Applying just transition approaches to the circular economy is important for identifying which countries, sectors, communities and workforces may be adversely affected by the process, as well as for developing policies and programmes to support those at risk of being left behind.[6] Therefore, as a preliminary study, the book is divided into topics and regions relative to these core pillars of the circular economy and the just transition to highlight and provide a tailored approach to the issues. Further, the aspiration is that the book will provide an impetus for further work on the circular economy and its inescapable relationship with the just transition. Overall, the topics aim to advance a global approach to issues which are at the core of the just transition today, namely:

1. Under the umbrella of *Reform*:
    - Gender issues with a focus on the initiatives taking place in parts of Africa and Brazil, as highlighted in Chapter 2;

[2] Vaibhav, H. 2025. Circular Economy: The Moder Hack for (Almost) Infinite Material Life. Available at: https://earth.org/circular-economy-the-modern-hack-for-almost-infinite-material-life/ (Accessed: 20 October 2025).

[3] Kirchherr, J. *et al.* 2023. Conceptualising the Circular Economy (Revisited): An Analysis of 221 definitions. *Resources, Conservation & Recycling, 194.* https://doi.org/10.1016/j.resconrec.2023.107001

[4] Ibid.

[5] Lakic, I. *et al.* 2023. A just energy transition takes a ecosystem. Boston Consulting Group. Available at: https://www.bcg.com/publications/2023/exploring-challenges-of-just-transition (Accessed: 20 October 2025).

[6] Schröder, P. 2020. Promoting a just transition to an inclusive circular economy. Chatam House. Available at: https://www.chathamhouse.org/2020/04/promoting-just-transition-inclusive-circular-economy-0/4-towards-just-circular-economy (Accessed: 22 October 2025).

   - The move to redevelop nuclear energy, showcasing the strategies of Europe in Chapter 3; and
   - How urban regeneration and design is changing the structure of cities and energy, with the introduction of energy communities, highlighting the cases of Roma and Brasilia in Chapter 4.
2. Valuable insights into *Reuse*:
   - By examining consumption patterns, as discussed in Chapter 5 through the fast fashion crisis and its impacts in Bangladesh, and
   - In Chapter 6 by highlighting the impacts of mining in Latin America – particularly in Peru.
3. An in-depth look into *Reduce*:
   - In Chapter 7, the accelerated renewable energy investment in the EU and Tunisia highlights the green hydrogen initiatives taking place to tackle GHG emissions in light of climate change.

Overall, the book places a first reliance on these three pillars that are pivotal aspects of the circular economy and which naturally lend themselves to correcting topical issues which are taking place in the just transition.

1.2 Framing What is 'Just'

Generally, a sustainable future requires major changes to how we power and operate our economies and will have inevitable ripple effects on livelihoods, but at the same time, transitioning from carbon-intensive industries to more sustainable ones can bring remarkable benefits.[7] Climate ambition, therefore, cannot be divorced from social and economic justice in that the process to decarbonise the economy should not be to the detriment of workers, communities, the environment, health, well-being and the economic security of states. To deliver the structural changes at the scale and speed required, carefully led state planning, proper regulations and the control of key policy and investment decisions must take place in a just way. Finding the equilibrium point is the heart of the just transition, in that the risks and benefits must be balanced with no one being left behind.

To do so, a targeted approach to a 'just' way must be initialised and encapsulate the principles of energy justice. With its five key principles of

[7] Gomez, C. *et al.* 2025. What is a 'Just Transition', and are countries really making progress? World Resources Institute. Available at: https://www.wri.org/insights/what-is-just-transition-tracking-progress. (Accessed: 20 October 2025).

justice (distributive, procedural, restorative, recognition and cosmopolitan),[8] energy justice can contribute to transformative change by promoting a green economy that is inclusive and non-discriminatory between generations by:

1. Reducing inequalities rather than exacerbating them;
2. Respecting human rights throughout the energy supply chain, with companies, governments, investors and civil society actors all having a vital role to play; and
3. Enabling alternate livelihoods for people and communities adversely affected by energy-related projects.

However, harmonising these is no small feat and includes many considerations, which include the integration of[9]:

1. The Well-being principle: where the green economy should be people-centred and prioritise investment and access to sustainable natural systems, infrastructure, knowledge and education for all people to prosper;
2. The Planetary boundaries principle: where the economy safeguards, restores and invests in nature by employing the precautionary principle to avoid loss of critical natural capital and breaching ecological limits, including climate stability;
3. The Efficiency and Sufficiency principle: where the economy should be geared to support sustainable consumption as well as sustainable production and the recognition that there must be a significant global shift to limit consumption of natural resources to physically sustainable levels; and
4. The Good Governance principle: The need for effective governance, which includes institutions that are integrated, collaborative and coherent. Therefore, requiring public participation, devolved decision-making and setting up ways that safely serve the interests of society.

[8] Heffron, R.J. 2023. Energy Justice-the triumvirate of tenets revisited and revised. *Journal of Energy & Natural Resources Law, 42*(2): 227–233. https://doi.org/10.1080/02646811.2023.2256593

[9] Partners for Inclusive green economy. 2019. Principles, priorities and pathways for inclusive green economies: Economic transformation to deliver the SDGs. Available at: https://www.greenpolicyplatform.org/research/principles-priorities-and-pathways-inclusive-green-economies-economic-transformation (Accessed: 20 October 2025).

Overall, therefore, the idea of 'just' encompasses various dimensions amounting to fairness, equity and inclusivity for everyone involved as the transition takes place in light of climate change. Procedurally, since law underpins society, the good governance principle should be at the helm of propelling what is just in the transition and deliver improved and more 'just outcomes' in society.

1.3 Future-proofing Society: Invoking Circular Economy Strategies Towards a Just Transition

Though its roots lie in business, the concept of future-proofing has never been more dire than it is today due to climate change. The concept of future-proofing is the process of anticipating the future and developing methods of minimising the negative impacts while taking advantage of the positive effects of shocks and stresses due to future events.[10] In the realm of sustainable and environmental issues, future-proofing is used to describe the ability of a design to resist the impact of potential climate change due to global warming.[11] If we effectively invoke the strategies under the circular economy, a twin benefit can occur by moving countries closer to a just transition and therefore future-proofing societies to last years to come.

Utilising the 3Rs approach here involves redesigning our systems not only for how we produce but also for how we consume and how we dispose to ensure that resources are continuously reused and regenerated over time. Firstly, via *reform*, sustainable governance is needed to future-proof society, a principle which has ties to the circular economy and the just transition alike. Sustainable governance is a holistic approach that integrates environmental, social and economic considerations to ensure long-term viability and well-being and involves creating policies, strategies and institutions that promote environmental sustainability, social equity and economic prosperity while preserving resources for future

[10] Rich, B. 2016. The Principles of Future-proofing: A broader understanding of resiliency in the historic built environment. Available at: https://www.researchgate.net/publication/333856761_The_Principles_of_Future-Proofing_A_Broader_Understanding_of_Resiliency_in_the_Historic_Built_Environment (Accessed: 20 October 2025).

[11] Ibid.

generations.[12] Secondly, *reuse* while targeting the way in which consumption takes place involves recycling and optimising efficiency and undoubtedly an approach like this would mitigate environmental damage and have the domino effect of building resilience. Thirdly, *reduce* manifests itself not only in reducing carbon emissions but also in optimising resource efficiency by employing renewable energy initiatives and implementing reduce-reuse-recycle policies to foster resource-conserving lifestyles.[13] By doing this, environmental justice is also triggered and allows for the mitigation of the impacts of climate change on the pressure ecosystems experience, a reduction in emissions and the protection of marginalised communities who would have been severely affected before.

1.4 Methodology

The supremacy of method in research should not be understated. Research methodology is an essential component of the research process. It provides a structured approach to conducting research, ensures the reliability and validity of research findings, facilitates communication and collaboration and helps to advance knowledge in a particular field.[14] Research is a fundamental driving force behind the advancement of knowledge and the progression of society and serves as a systematic exploration aimed at uncovering new insights, challenging existing paradigms and solving complex problems.[15] The ideas explored in this book are presented by early career researchers who all believe that energy justice is a powerful

[12] Gireesan. K. 2024. Sustainable Governance. In: *Democracy, Leadership and Governance – Application of Artificial Intelligence*. Palgrave Macmillan, Singapore. https://doi.org/10.1007/978-981-99-7735-2_6

[13] Velenturf, A.P.M. and Purnell, P. 2021. Principles for a sustainable circular economy. *Sustainable Production and Consumption*, *27*: 1437–1457. Available at: https://www.sciencedirect.com/science/article/pii/S2352550921000567 (Accessed: 22 October 2025).

[14] Bahishti, Adam. 2022. The Vital role of research methodology in addressing the research questions. *International Journal of Methodology, 1*. https://doi.org/10.21467/ijm.1.1.2016.29

[15] Islam, M. 2023. The Importance of Research in the Advancement of Knowledge and Society. *International Research Journal of Basic and Clinical Studies, 8*(4): 1–3. https://doi.org/ 10.14303/irjbcs.2023.49

tool which can be used to navigate and correct the issues taking place in the world today.

To explore and navigate these issues, the authors have employed a mix of both quantitative and qualitative methods. The primary advantage of quantitative research is that statistical evidence can be used to enhance confidence in a knowledge claim, rule out rival explanations and enhance confidence in knowledge claims.[16] In this book, the authors have employed the use of numerical data and statistical analysis to buffer the qualitative research done. Statistical techniques allow researchers to make generalisations about a wider population event or objects based on the sample data they collect to address specific quantitative research enquiries, thus allowing them to make sense of complex questions and gain deeper insights from their data.[17] In qualitative research, methods commonly used in qualitative practice include but are not limited to ethnography, autoethnography, duo ethnography, narrative inquiry, in-depth interview, semi-structured interview, focus group interview, oral history, document analysis, content analysis, historical-comparative methods, poetic inquiry, audiovisual methods, visual methods, photovoice, case study, multiple case study, discourse analysis, conversation analysis,[18] etc. thus meaning that qualitative research includes a range of tools for data collection.

In this book the methods range from a case study, multiple case studies, visual methods, content and document analysis as well as historical analyses. Further, there is also a fusion of methods which aids in increasing study credibility.

1.5 Organisation of the Text

This book is organised into five sections and consists of nine chapters. *Part I* of the book situates the research by providing the aim and introduces the 3Rs, which will form the basis of the text. Part I stresses that the book is a

[16] Levine, T. Quantitative Social Science Methods of Inquiry. In: Knapp, M.L. and Daly, J.A. (eds). *The Sage Handbook of Interpersonal Communication*. 4th Edition. Sage Publications.

[17] Singh, D.P. and Jassi, J. Explorign the significance of statistics in the research: A comprehensive overview section A- Research paper. *European Chemical Bulletin, 12*: 2089–2102. https://doi.org/10.31838/ecb/2023.12.s2.262

[18] Leavy, P. 2014. Introduction to the Oxford Handbook of Qualitative Research, Second Edition. In: Leavy, P (ed.) *The Oxford Handbook of Qualitative Research*. Oxford University Press.

preliminary study of the issues which are currently taking place in the just transition and how we can move closer to solving the issues utilising the 3R circular economy principles. *Part II* highlights what is being done at a communal level in Brazil and Africa by women-led organisations and at a state level in the EU, Italy and Brazil to reform or transform the respective societies as they navigate the transition.

Part III then examines the notion of reuse via consumption patterns and how we can refocus it on efficiency and sustainability by looking at the fast fashion crisis in Bangladesh and the effects of mining practices in Latin America. *Part IV* shifts the focus to a topical approach to reducing GHG emissions, namely the use of green hydrogen, and collaboratively investigates the strategies being employed by the EU and Tunisia from a procedural justice lens. *Part V* concludes the book in two chapters, calling for the need to promote further research in the areas of energy justice, the just transition and the circular economy.

1.6 Conclusion

We would like to reiterate that this book is only a preliminary study of three principles of the circular economy, and it is hoped that this book will give confidence to researchers and steer more research in the area. While the title of the book is targeted to be a global perspective, we believe that the regions and topics chosen are crucial parts of the just transition which can benefit from the employment of circular economy strategies. A just transition is not merely adjunct to climate action, in fact it is the nucleus since it drives the focus for a more moral, social and economic imperative. By integrating the three principles of the circular economy, which emphasise the need to restructure society (Reform), maximise resource efficiency sustainably (Reuse), reducing waste and promoting climate ambitions (Reduce) becomes a straightforward pathway to achieve the just transition. Ultimately the transition to a sustainable future will only be successful if it is both just and circular, advancing side by side to ensure no one is left behind.

And now on to the rest of the book!

PART II

Reform, Restructure and Reorganize to Transform Society

CHAPTER 2

Whose Energy Justice? Situated Knowledges from Women-led Grass Roots 'Organising' in the Global South

Beatriz Arnal-Calvo, Carmen Freed Huici, and Kathlen Schneider

Abstract This chapter advocates for the inclusion of silenced voices in energy justice research and policy. Drawing on and contributing to decolonial, intersectional and feminist theory, it problematises who is included and excluded and whose voices, knowledges and experiences matter in energy justice debates. People's experiences of and contributions to energy transitions shape and are simultaneously shaped by gender relations at the intersection of class, race, ethnicity and geographical location. Indigenous women and women of colour in the Global South, often at the front line of advancing energy just futures, experience the intersection of these forms of violence and marginalisation in particularly acute ways. Ignoring these

B. Arnal-Calvo
School of Humanities and Social Science, University of Brighton, United Kingdom
e-mail: B.ArnalCalvo@brighton.ac.uk

C. Freed Huici
The Bartlett, University College London, United Kingdom

K. Schneider
Science Policy Research Unit (SPRU), University of Sussex, Falmer, United Kingdom

A. Phillips, L. Kaschny (eds.), *Global Perspectives on the Circular Economy in the Just Transition*, Just Transitions,
https://doi.org/10.1007/978-3-032-18342-2_2

intersecting realities and simplifying women's roles risks perpetuating underlying inequalities while at the same time neglecting the many ways women contribute to and lead a just energy transformation for all. Feminist, intersectional and decolonial inquiry is best positioned to challenge structural power dynamics at the core of energy justice scholarship and policymaking, specifically around epistemic injustice and the exclusion of lived experiences, embodied knowledges and situated demands from women in the Global South. By interrogating who the political subjects and knowledge producers in energy justice are, this chapter amplifies contributions from two women-led, grassroots organising in the Global South – *Arpilleras* in Brazil and WoMin in the African continent – in their own terms. These grassroots 'organising' in the Global South – a concept used to encapsulate the diverse ways in which people organise, mobilise and gather in demand for justice – provide tangible opportunities to advance more inclusive and transformative approaches to energy justice research and policy. Further, they contribute to a fundamental rethinking of how knowledge is valued, recognising that expertise also exists in communities, traditions and lived practices.

Keywords Energy justice • Epistemic justice • Feminist methodologies • Intersectionality • Women-led • Grassroots organising • Global South

2.1 Introduction

This chapter advocates for the inclusion of silenced voices across different lines of discrimination, namely gender, class, race, ethnicity and geographical location. This collective work followed a thorough discussion and self-reflection on power, discrimination and positionality among the authors and the ethical dilemmas of excluding, speaking about or on behalf of people from unfamiliar realities. In this sense, the three chapter's authors want to acknowledge their multiple identities as women researchers based in the UK with personal and/or professional experience in the Global South (mainly in Mozambique, Chile and Brazil). To navigate the intricacies inherent in critical scholarship, this chapter problematises who is included and excluded and whose voices, knowledges and experiences matter in energy justice. Or, in Spivak's words, 'Can the subaltern speak?'[1] In particular, this chapter seeks to answer the following research questions:

[1] Spivak, G. C. 1988. 'Can the Subaltern Speak?' in C. Nelson and L. Grossberg (eds), *Marxism and the Interpretation of Culture*. University of Illinois Press, as in original paper.

1. how women-led organising from the Global South articulate their actions in demand of energy justice;
2. how their lived experiences, embodied knowledge and situated demands inform those actions;
3. whether and how they contribute to existing but narrow hegemonic energy justice frameworks, for instance, by reproducing, contesting, resisting, amplifying or complementing them and
4. how their contributions can advance more inclusive and transformative energy justice research and policies.

Following this introduction, the next section reviews normative energy justice frameworks and their (lack of) inclusion of decolonial, intersectional and feminist approaches. It focuses on epistemic justice and the recognition of the pluriversality of lived experiences, embodied knowledges and situated demands from the ground, from which this chapter progresses towards a more inclusive, just and transformative energy justice framework. After a methodological section, this chapter introduces two women-led, grassroots organising in the Global South – *Arpilleras* in Brazil and WoMin in the African continent – and discusses what they bring new to the energy justice debate. By bridging energy justice frameworks along with epistemic justice and critical intersectional insights, this chapter amplifies what they experience, know and demand in their own terms. The conclusion highlights a series of final remarks and policy recommendations on how to acknowledge and include historically marginalised voices in energy justice transformations, so the question 'whose energy justice?' is no longer relevant.

2.2 Theoretical Framework: Critical Perspectives, Intersectionality and Epistemic Justice

2.2.1 Energy Justice

The concept of energy justice integrates human rights across the entire energy lifecycle, from extraction to production, operation, consumption and waste management, including decommissioning.[2] Various approaches

[2] McCauley, D. and Heffron, R. 2018. Just Transition: Integrating Climate, Energy and Environmental Justice. *Energy Policy*, *119*: 1–7. https://doi.org/10.1016/j.enpol.2018.04.014; Heffron, R.J. 2023. Energy Justice–the Triumvirate of Tenets Revisited and Revised.

to energy justice frameworks are found in the literature, most of them built around three core principles: distributive justice, procedural justice and recognition justice.[3] Distributive justice refers to the distribution of goods and bads (also seen as resources and harms);[4] procedural justice includes access to information, participation in decision-making and the ability to influence outcomes;[5] and recognition justice addresses who is given respect and who is not valued.[6] Over time, scholars like Heffron[7] have argued for an expanded framework to include restorative and cosmopolitan justice, recognising the need to address the global and intergenerational dimensions of energy justice: restorative justice focuses on the needs of victims and communities harmed, emphasising the obligation to repair the damage already caused, whether intentional or unintended,[8] while cosmopolitan justice extends the scope of justice beyond national borders, emphasising the interconnectedness of energy policies and their effects on a global scale, particularly considering international commitments such as the Paris Agreement.[9] In practice, these deeply intercon-

Journal of Energy & Natural Resources Law, *42*(2): 227–233. https://doi.org/10.1080/02646811.2023.2256593; Heffron, R. J. and McCauley, D. 2018. What Is the 'Just Transition'? *Geoforum*, *88*:74–77. https://doi.org/10.1016/j.geoforum.2017.11.016

[3] Wang, X. and Lo, K. 2021. Just Transition: A Conceptual Review. *Energy Research and Social Science, 82*. https://doi.org/10.1016/j.erss.2021.102291

[4] Dynge, M. and Cali, U. 2025. Distributive Energy Justice in Local Electricity Markets: Assessing the Performance of Fairness Indicators. *Applied Energy*, 384. https://doi.org/10.1016/j.apenergy.2025.125463

[5] Hoesch, K. *et al.* 2025. What to Expect When You're Expecting Engagement: Delivering Procedural Justice in Large-Scale Solar Energy Deployment. *Energy Research & Social Science*, 120. https://doi.org/10.1016/j.erss.2024.103893

[6] Williams, S. and Doyon, A. 2019. Justice in Energy Transitions. *Environmental Innovation and Societal Transitions,* 31. https://doi.org/10.1016/j.eist.2018.12.001; Van Uffelen, N. 2022. Revisiting Recognition in Energy Justice. *Energy Research & Social Science, 92*. https://doi.org/10.1016/j.erss.2022.102764

[7] Heffron, R.J. 2023. Energy Justice–the Triumvirate of Tenets Revisited and Revised. *Journal of Energy & Natural Resources Law*, *42*(2): 227–233. https://doi.org/10.1080/02646811.2023.2256593

[8] Hazrati, M. and Heffron, R.J. 2021.Conceptualising Restorative Justice in the Energy Transition: Changing the Perspectives of Fossil Fuels. *Energy Research and Social Science, 78*. https://doi.org/10.1016/j.erss.2021.102115

[9] Heffron, R.J. 2023. Energy Justice–the Triumvirate of Tenets Revisited and Revised. *Journal of Energy & Natural Resources Law*, *42*(2): 227–233. https://doi.org/10.1080/02646811.2023.2256593; Phillips, A. 2024. The Quest for Cosmopolitan Justice in the Energy Transition in Caribbean Small Island Developing States. In: Heffron, R.J. and de Fontenelle, L. (eds). T*he Power of Energy Justice and the Social Contract*. Just Transitions.

nected and mutually reinforcing principles[10] can be applied to evaluate and guide energy projects and policies, ensuring that they contribute to a just and equitable energy transition.

Despite the rising importance of these frameworks, much of the scholarship on energy justice and policy development in just transitions has been concentrated in the Global North. Studies have largely reflected Western and masculinist assumptions, without adequately interrogating the structural inequalities and power relations underpinning energy systems[11] and leaving limited space for the diverse experiences and epistemologies from the Global South.[12] Critical perspectives, however, challenge these universalist understandings of justice, advocating for a more plural and situated approach rooted in cultural, territorial, embodied and historical realities.

2.2.2 Decolonial Theory

Scholars from Latin America foreground the colonial legacies and persistent power asymmetries that shape energy systems and injustices.[13]

Palgrave Macmillan, Cham. https://doi.org/10.1007/978-3-031-46282-5_21; Sovacool, B.K. *et al.* 2016. Energy Decisions Reframed as Justice and Ethical Concerns. *Nature Energy, 1.* https://doi.org/10.1038/nenergy.2016.24

[10] Heffron, R.J. 2023. Energy Justice–the Triumvirate of Tenets Revisited and Revised. *Journal of Energy & Natural Resources Law*, *42*(2): 227–233. https://doi.org/10.1080/02646811.2023.2256593; Wang, X. and Lo, K. (2021). Just Transition: A Conceptual Review. *Energy Research and Social Science, 82.* https://doi.org/10.1016/j.erss.2021.102291; Hazrati, M. and Heffron, R.J. 2021. Conceptualising Restorative Justice in the Energy Transition: Changing the Perspectives of Fossil Fuels. *Energy Research and Social Science*, 78. https://doi.org/10.1016/j.erss.2021.102115

[11] Wood, N. 2023. Problematising Energy Justice: Towards Conceptual and Normative Alignment. *Energy Research & Social Science, 97.* https://doi.org/10.1016/j.erss.2023.102993; Van Uffelen, N.; Taebi, B. and Pesch, U. 2024. Revisiting the Energy Justice Framework: Doing Justice to Normative Uncertainties. *Renewable and Sustainable Energy Review, 189, Part A.* https://doi.org/10.1016/j.rser.2023.113974; Tornel, C. 2022. Decolonizing Energy Justice from the Ground up: Political Ecology, Ontology, and Energy Landscapes. *Progress in Human Geography*, *47*(1). https://doi.org/10.1177/03091325221132561

[12] Lacey-Barnacle, M.; Robison, R. and Foulds, C. 2020. Energy Justice in the Developing World: A Review of Theoretical Frameworks, Key Research Themes and Policy Implications. *Energy for Sustainable Development*, *55*: 122–138. https://doi.org/10.1016/j.esd.2020.01.010; Sovacool, B.K. *et al.* (2017). New Frontiers and Conceptual Frameworks for Energy Justice. *Energy Policy, 105*: 677–691. https://doi.org/10.1016/j.enpol.2017.03.005

[13] De Freitas, A. 2019. Decolonial Epistemologies for Energy Planning in Brazil. PhD Thesis, University of Westminster School of Law. https://doi.org/10.34737/v3497; Mendes, E. 2024. Just Energy Transition Policies at the Subnational Level: A Critical and

The term 'coloniality' refers to the long-term effects of colonialism, which extend beyond the formal political end of empires[14] and continue to shape global power dynamics and knowledge hierarchies, particularly impacting the Global South. In this context, decoloniality, as a critical theoretical framework, emerged primarily from Latin America and the Caribbean, where the experiences of colonialism are particularly profound and enduring.[15] Scholars from this region played a pivotal role in articulating the concept and its impact on various aspects of life, from economics and politics to culture and epistemology. Quijano's concept of 'coloniality of power', a cornerstone of decolonial theory,[16] shows how colonialism established a global racial hierarchy that continues to influence power relations in the contemporary world, for instance by shaping global knowledge production.[17] Mignolo's work focuses on how knowledge is imposed globally by Europe, marginalising alternative ways of knowing and understanding the world.[18] He emphasises the need to decolonise knowledge production by creating space for diverse epistemologies and challenging the dominance of Western intellectual traditions.[19] The decolonial perspective has

Propositional Analysis Based on the Just Energy Transition Plan of the State of Ceará [Políticas de Transição Energética Justa Em Nível Subnacional: Uma Análise Crítico-Propositiva a Partir Do Plano de Transição Energética Justa Do Estado Do Ceará, In Portuguese]. Master's Dissertation, Universidade Federal do Ceará. http://repositorio.ufc.br/handle/riufc/76486; Roysen, R. *et al.* (2025). Building a World-Ship: A Decolonial Approach to Just Energy Transitions. *Human Geography*. https://doi.org/10.1177/19427786251326605; Tornel, C. 2022. Decolonizing Energy Justice from the Ground up: Political Ecology, Ontology, and Energy Landscapes. *Progress in Human Geography*, *47*(1). https://doi.org/10.1177/03091325221132561

[14] Fúnez-Flores, J. *et al.* (2022). Decolonial Discourses and Practices: Geopolitical Contexts, Intellectual Genealogies, and Situated Pedagogies. *Educational Studies, 58*(5–6): 596–619. https://doi.org/10.1080/00131946.2022.2132393

[15] Fúnez-Flores, J.; Díaz Beltrán, A.C. and Jupp, J. 2022. Decolonial Discourses and Practices: Geopolitical Contexts, Intellectual Genealogies, and Situated Pedagogies. *Educational Studies, 58*(5–6): 596–619. https://doi.org/10.1080/00131946.2022.2132393

[16] Ibid.

[17] Couldry, N. and Mejias, U.A. 2021. The Decolonial Turn in Data and Technology Research: What Is at Stake and Where Is It Heading? *Information, Communication & Society, 26*(4):786–802. https://doi.org/10.1080/1369118X.2021.1986102

[18] Fúnez-Flores, J.; Díaz Beltrán, A.C. and Jupp, J. 2022. Decolonial Discourses and Practices: Geopolitical Contexts, Intellectual Genealogies, and Situated Pedagogies. *Educational Studies, 58* (5–6): 596–619. https://doi.org/10.1080/00131946.2022.2132393

[19] Haneef, M.A.2003. Decolonizing Methodologies: Research and Indigenous Peoples by Linda Tuhiwai Smith (London, UK: Zed Books Ltd. University of Otago Press, 1999:

since spread globally, finding resonance in various contexts and informing critical analyses of colonialism's lasting impact on communities worldwide.[20]

Ultimately, decolonial theory highlights how colonial power structures not only imposed political and economic control but also shaped knowledge systems, privileging European ways of knowing while marginalising Indigenous and other non-Western knowledge systems.[21] Accordingly, decolonial thinking challenges the enduring mechanisms of power by which colonial legacies are perpetuated,[22] including how coloniality shapes knowledge production, social relations and subjectivity itself. In this sense, a decolonial perspective to energy justice problematises the systematic exclusion of the Global South, which shares common experiences of colonisation, from the dominant narrative of energy justice.[23] As scholars argue, concepts like energy and justice are not universally understood or evident across the globe.[24]

208 pages. *American Journal of Islam and Society, 20*(3–4): 188–191. https://doi.org/10.35632/ajis.v20i3-4.1836

[20] Ibrahima, A.B. and Mattaini, M.A. 2018. Social Work in Africa: Decolonizing Methodologies and Approaches. *International Social Work, 62*(2): 799–813. https://doi.org/10.1177/0020872817742702

[21] Couldry, N. and Mejias, U.A. 2021. The Decolonial Turn in Data and Technology Research: What Is at Stake and Where Is It Heading? *Information, Communication & Society, 26*(4): 786–802. https://doi.org/10.1080/1369118X.2021.1986102

[22] Radcliffe, S.A. 2017. Decolonising Geographical Knowledges. *Transactions of the Institute of British Geographers, 42*(3): 329–333. https://doi.org/10.1111/tran.12195; Shakun, N. *et al.* 2024. Postcolonial Perspectives in the 21st Century: A Critical Analysis of Modern Theories and Methodologies. *Revista Amazonia Investiga, 13*(78):194–208. https://doi.org/10.34069/AI/2024.78.06.16

[23] Kumar, K., Höffken, J. and Pols, A. (Eds.). 2021. *Dilemmas of Energy Transitions in the Global South: Balancing Urgency and Justice* (1st ed.) Routledge. https://doi.org/10.4324/9780367486457; Abe, O. and Azubike, V. 2024. (Re)Examining the Intersection between Energy Justice and Energy Transition in Africa. *Journal of Energy & Natural Resources Law, 42*(3): 279–299. https://doi.org/10.1080/02646811.2024.2356988; Monyei, C.G. *et al.* 2018. Examining Energy Sufficiency and Energy Mobility in the Global South through the Energy Justice Framework. *Energy Policy, 119*: 68–76. https://doi.org/10.1016/j.enpol.2018.04.026; Apergi, M. *et al.* 2024. An Energy Justice Index for the Energy Transition in the Global South. *Renewable and Sustainable Energy Reviews, 192.* https://doi.org/10.1016/j.rser.2023.114238; Nsafon, B.E.K. *et al.* 2023. The Justice and Policy Implications of Clean Energy Transition in Africa. *Frontiers in Environmental Science, 11.* https://doi.org/10.3389/fenvs.2023.1089391

[24] Valkenburg, G. 2024. Energy Justice as Epistemic Justice. *Ethics, Policy & Environment,* 1–22. https://doi.org/10.1080/21550085.2024.2418789

2.2.3 *Intersectional Feminist Theory*

Feminist scholars have long challenged the illusory nature of academic objectivity, neutrality and detachment. Haraway's[25] concept of 'situated knowledges' argues that all knowledge is shaped by the specific social, historical and embodied circumstances of the knower. Similarly, for Shepherd,[26] knowledge production is always subjective, partial and relational, based on relations of power and identity formation. Furthermore, academic authority and expertise are profoundly gendered as gender identities intersect with scholarly legitimacy.[27] Ackerly and True,[28] correspondingly, argue that underlying power dynamics in knowledge production processes determine what counts as legitimate. On the other hand, intersectionality, as originated by Crenshaw[29] and further developed by Black feminist scholarship, provides a conceptual framework to examine how overlapping gendered, racialised and class-based identities combine to disproportionally marginalise certain communities. Intersectional feminist scholarship highlights how interconnected systems of power and exclusion – namely class, race, ethnicity, religious affiliation, sexual orientation and disability – structure people's access to, participation in and impact from energy systems.[30] As a result, women, especially indigenous women and women of colour from the Global South, experience particularly acute forms of violence and marginalisation at these intersections,[31] while their multiple situated knowledges and lived experiences are systematically disregarded.

[25] Haraway, D. 1988. Situated Knowledges: The Science Question in Feminism and the Privilege of Partial Perspective. *Feminist Studies, 14*(3): 575–599. https://doi.org/10.2307/3178066

[26] Shepherd, L.J. 2016. Research as Gendered Intervention: Feminist Research Ethics and the Self in the Research Encounter. *Critica Contemporánea: Revista de Teoría Política, 6*: 1–15.

[27] Ibid.

[28] Ackerly, B.A. and True, J. 2019. *Doing Feminist Research in Political and Social Science* (2nd edn, Macmillan International 2020).

[29] Crenshaw, K. 1991. Mapping the Margins: Intersectionality, Identity Politics, and Violence against Women of Color. *Stanford Law Review, 43*(6): 1241–1299. https://doi.org/10.2307/1229039

[30] Robinson, C., Simcock, N. and Petrova, S. (2023). Energy Justice and Gender. In *Handbook on Energy Justice*. Cheltenham, UK: Edward Elgar Publishing. https://doi.org/10.4337/9781839102967.00020

[31] Ibid.

Because energy systems are not neutral but embedded in and shaped by historical and ongoing relations of domination,[32] power-blind energy justice scholarship and policymaking risk perpetuating the very inequalities they seek to address.[33] Conversely, intersectional feminist epistemologies ensure more transformative, inclusive and context-specific approaches by addressing not only distributive and procedural injustice but also deeper structural and epistemic exclusions that determine what forms of knowledge are ignored and whose worldviews are subjugated.

2.2.4 Epistemic (In)Justice

Defined by Fricker,[34] the concept of epistemic injustice refers to the wrongs committed against someone as a subject of knowledge. Fricker identified two kinds of injustices: testimonial injustice occurs when a person is given too little credibility due to identity prejudice on the hearer's part, often related to intersectional racial, class and gender identities. By its turn, hermeneutical injustice happens when people lack the conceptual resources to articulate or interpret their own experiences, placing them at an unfair disadvantage.[35] In this sense, epistemic justice seeks to ensure

[32] Sovacool, B.K. *et al.* 2023. Pluralizing Energy Justice: Incorporating Feminist, Anti-Racist, Indigenous, and Postcolonial Perspectives. *Energy Research & Social Science, 97*. https://doi.org/10.1016/j.erss.2023.102996; Listo, R. 2018. Gender Myths in Energy Poverty Literature: A Critical Discourse Analysis. *Energy Research and Social Science, 38*: 9–18. https://doi.org/10.1016/j.erss.2018.01.010

[33] Bell, S.E.; Daggett, C. and Labuski, C. 2020. Toward Feminist Energy Systems: Why Adding Women and Solar Panels Is Not Enough☆. *Energy Research & Social Science, 68*. https://doi.org/10.1016/j.erss.2020.101557; Götzmann, N. and Dicalou, M. 2025. Towards a Feminist Energy Justice Framework. *Business and Human Rights Journal, 10*(1): 56–78.doi:10.1017/bhj.2025.13; Robinson, C., Simcock, N. & Petrova, S. 2023. Energy Justice and Gender. In *Handbook on Energy Justice*. Cheltenham, UK: Edward Elgar Publishing. https://doi.org/10.4337/9781839102967.00020; Sovacool, B.K. *et al.* 2023. Pluralizing Energy Justice: Incorporating Feminist, Anti-Racist, Indigenous, and Postcolonial Perspectives. *Energy Research & Social Science, 97*. https://doi.org/10.1016/j.erss.2023.102996; Walk, P. 2024. From Parity to Degrowth: Unpacking Narratives of a Gender-Just Transition. *Energy Research & Social Science, 112*. https://doi.org/10.1016/j.erss.2024.103513

[34] Fricker, M. 2007. *Epistemic Injustice: Power and the Ethics of Knowing*. Oxford Academic. https://doi.org/10.1093/acprof:oso/9780198237907.001.0001

[35] Cummings, S. *et al.* 2023. Doing Epistemic Justice in Sustainable Development: Applying the Philosophical Concept of Epistemic Injustice to the Real World. *Sustainable Development, 31*(3):1965–1977. https://doi.org/10.1002/sd.2497; Valkenburg, G. 2024. Energy

that everyone contributes to knowledge and that their testimonies and experiences are valued impartially.

In the context of energy justice, Valkenburg[36] argues that epistemic justice goes beyond traditional frameworks that focus on the distribution of benefits and procedural participation. Instead, epistemic justice provides a critical lens for redefining all the other pillars of energy justice, as all depend on whose knowledge is counted as credible and valuable.[37] Its purpose is to ensure that local, experiential and ancestral knowledge is not dismissed as irrational but considered as fundamental as scientific or technical knowledge in decision-making. Furthermore, Valkenburg[38] encourages refocusing the energy justice agenda from simply including diverse voices to actively transforming the structures of knowledge recognition, exchange and action. The question, then, is not only about including diverse knowledge systems and perspectives about energy and justice but about how that inclusion is done.[39] Epistemic justice, thereby, specifically addresses the devaluation and exclusion of particular perspectives in the process of knowledge production. This entails, but exceeds, bridging epistemic differences, acknowledging diverse sources of knowledge and ensuring that marginalised communities, often the most affected by energy projects and policies, are heard and contribute to decision-making.[40] Further, epistemic justice advocates for the dismantlement of colonial forms of categorisation, quantification and standardisation that determine whose knowledge counts and whose futures matter in energy governance.[41] Energy justice cannot be achieved without confronting the deep structures of knowing and being that underpin colonial modernity.[42] Within this account, energy transitions must not only redistribute resources or improve participation but also transform

Justice as Epistemic Justice. *Ethics, Policy & Environment*, 1–22. https://doi.org/10.1080/21550085.2024.2418789

[36] Valkenburg, G. 2024. Energy Justice as Epistemic Justice. *Ethics, Policy & Environment*, 1–22. https://doi.org/10.1080/21550085.2024.2418789

[37] Ibid.

[38] Ibid.

[39] Ibid.

[40] Valkenburg, G. 2024. Energy Justice as Epistemic Justice. *Ethics, Policy & Environment*, 1–22. https://doi.org/10.1080/21550085.2024.2418789

[41] Arora, S.; Ghosh, B. and Stirling, A. 2025. *Decolonising Innovation for Pluriversal Justice in Sustainability Transitions.* Social Science Research Network.

[42] Ibid.

how knowledge is produced, whose knowledge is valued and which worldviews inform just energy futures.

What the above demonstrates is that for energy justice frameworks to be truly just and transformative, they ought to acknowledge the diverse experiences, knowledges and identities of historically marginalised communities and individuals and how interlocking power systems shape their very engagement in energy transitions. By centring the lived experiences, embodied knowledges and situated demands of women-led, non-governmental organising from the Global South, this chapter challenges the assumption that there is such a thing as a unique, universal and objective encounter with energy systems, technologies and processes. Instead, it acknowledges and values the plurality of sources and types of knowledge, particularly those context-specific and grounded in everyday experiences, often marginalised. In doing so, it offers significant contributions to the energy justice literature while challenging structural power dynamics at the core of energy justice research and policymaking.

2.3 Methodology, Analytical Dimensions and Research Questions

Drawing on the above, this chapter critically examines and challenges dominant energy justice frameworks. It offers a critical, comprehensive and transformative framework informed by the intersecting lived experiences, embodied knowledges and situated demands of those historically marginalised, which includes the following analytical dimensions:

1. *Epistemic justice and the recognition of plural knowledges*: this chapter recentres diverse, situated and intersectional knowledges, especially those historically marginalised by dominant Western, technocratic and patriarchal frameworks within energy justice, critically contesting whose knowledges are included and excluded in energy decision-making and policy processes.
2. *Intersectionality*: gender, in intersection with race, class, indigeneity, colonial history and other axes of power, shapes specific experiences of energy (in)justice. An intersectional lens ensures that energy processes address, without reproducing, complex and overlapping forms of oppression and privilege within energy systems.
3. *Structural power*: this chapter draws attention to how existing energy institutions, policies and technologies reproduce or challenge

underlying social hierarchies and structural inequalities, thus challenging power relations that determine access, control and profits from the energy sector.

4. *Sovereignty*: individuals and communities have the right to control their energy resources, production and consumption and to make decisions about their energy futures independently and sustainably. This chapter amplifies demands from participatory, democratic and collectively led approaches to energy futures so historically marginalised groups are not 'tokenised', but their lived experiences, embodied knowledge and situated demands effectively influence energy decisions and outcomes.
5. *Intersecting problems and root causes*: unlike technocratic and apolitical analyses that only address the immediate symptoms of injustice and discrimination, this chapter looks at the intertwined root causes of inequality and exclusion.
6. *Transformative solutions*: instead of superficial solutions that do not advance structural change, this chapter fosters the radical transformation of hegemonic energy rules, norms, values and frameworks that determine whose interests and knowledge are prioritised.
7. *Care ethics and the reproduction of life*: decolonial, feminist and intersectional theories favour energy systems that support care praxes contributing to the reproduction and flourishing of life.
8. *Accountability*: different stakeholders have common but differentiated rights and responsibilities in energy transitions; hence, they must be held accountable when energy systems are not responsive to the lived experiences, embodied knowledges and situated demands of historically marginalised groups.

By centring the above analytical dimensions, this novel energy justice framework draws attention to and amplifies the lived experiences, embodied knowledges and situated demands around energy justice from two women-led, grassroots organising in the Global South: the Movement of People Affected by Dams (MAB)'s *Arpilleras Project* in Brazil and *WoMin Pan-African Ecofeminist Alliance* (WoMin) in Africa. The term 'organising' is employed here to encapsulate the diverse ways in which people organise, mobilise and gather in demand for justice. Both case studies exemplify women-led, community-centred, decolonial, grassroot approaches to energy and gender justice from the Global South. The next section analyses their publicly available materials (e.g. websites, memos, speeches) concerning the aforementioned analytical dimensions while addressing four research questions:

1. how women-led organising from the Global South articulate their actions in demand of energy justice;
2. how their lived experiences, embodied knowledge and situated demands inform those actions;
3. whether and how they contribute to existing but narrow hegemonic energy justice frameworks, for instance by reproducing, contesting, resisting, amplifying or complementing them and
4. how their contributions can advance more inclusive and transformative energy justice research and policies.

2.4 Case Studies

The following case studies present two examples of women-led, grassroots organising for energy justice in the Global South. While grounded in different political, cultural and geographical contexts, both the *Arpilleras Project by the Movement of People Affected by Dams* (MAB) in Brazil and the *WoMin Pan-African Ecofeminist Alliance* show how women are resisting harmful energy systems, reclaiming their voices and proposing alternative development pathways. Table 2.1 offers a comparative summary of the two cases, highlighting their main characteristics.

2.4.1 *Arpilleras:* ***Stitching Stories of Resistance (Brazil)***

The *Movement of People Affected by Dams* (*Movimento dos Atingidos por Barragens* – MAB; Fig. 2.1) is a national, popular and independent movement that began in the 1980s when local communities across Brazil came together to resist big dam projects.[43] These dams were built to generate electricity, often forcing people from their homes, destroying rivers and lands and causing deep social and environmental harm. Over time, these local groups joined forces and became a national movement in 1991.[44] MAB fights for the rights of people who are affected by dams throughout

[43] Movimento Dos Atingidos Por Barragens. 'Quem Somos.' Available at: https://mab.org.br/quem-somos/ (Accessed: 29 June 2025).

[44] Movimento Dos Atingidos Por Barragens. 'Quem Somos.' Available at: https://mab.org.br/quem-somos/ (Accessed: 29 June 2025).

Table 2.1 Comparative summary table of the two case studies.

Organisation	*MAB – Arpilleras Project*	*WoMin*
Organisational classification	Independent national social movement	Pan-African ecofeminist alliance
Geographical scope	Brazil-wide	Specific sites across the African continent as well as at the sub-regional and regional levels.
Year founded	MAB emerged in the late 1980s. The Arpillera Project in 2013	2013
Goals	MAB fights for the realisation of human rights for the affected population by violations on dam constructions. The Arpillera Project aims to denounce the gendered impacts of dam-related injustices through collective textile art, while fostering women's political organisation, memory and resistance.	• Make visible and publicise the impacts of extractive industries on peasant and working-class women; • Support women's organising, movement-building and solidarity; • Advance an African post-extractivist, ecologically just, women-centred alternative to the dominant, destructive model of development.
Governance	Horizontal and participatory governance, based on the autonomy of affected communities and organised through local committees, regional meetings and national congresses.	Alliance managed by a team of activists and researchers from across Africa; governed through networked collaboration
Funding sources	Voluntary contributions, project-based grants and advocacy for public funding mechanisms	Mix of international grants and partnerships with feminist and philanthropic foundations
Main Actions	MAB operates across five main areas: energy, water, human rights, dams and the Amazon. Arpillera Project main actions include organising self-led sewing workshops across Brazil, producing collective textile narratives that document lived experiences of injustice and publicly exhibiting the *arpilleras* as tools for political education, mobilisation and feminist resistance.	• Energy and climate justice; • Consent Rights for Women and Their Communities; • Addressing Violence Against Women; • Advancing African Ecofeminist Development Alternatives.

Source (2025): Created by Arnal-Calvo, Beatriz; Freed Huici, Carmen; Schneider, Kathlen.

Fig. 2.1 Demonstrations from the Movement of People Affected by Dams (MAB). Source: Brasil de Fato (2017, 2025). Brasil de Fato. 2025. 'MAB Realiza Atos Em Defesa de Comunidades Atingidas Por Barragens Pelo Brasil.' Available at: https://www.brasildefato.com.br/2025/03/14/mab-realiza-atos-em-defesa-de-comunidades-atingidas-por-barragens-pelo-brasil/ (Accessed: 29 June 2025); Brasil de Fato. 2017. 'De Atingidos Por Barragens a Defensores Dos Rios, Da Água e Da Vida.' Available at: https://www.brasildefato.com.br/2017/03/16/de-atingidos-por-barragens-a-defensores-dos-rios-da-agua-e-da-vida/ (Accessed: 10 June 2025).

the project lifecycle. It is open to everyone, regardless of skin colour, gender, religion, or background. One of its main goals is to question and change Brazil's energy model, which often benefits big companies and harms local communities. MAB believes in solidarity, community organising and building strength from the ground up.

From the beginning, women have played a vital role in MAB. They helped organise actions, took care of the community and led many of the struggles. Over the years, MAB women in different regions of the country started talking more about how dams affect their lives in specific ways – how they lose their homes and work, how they were left out of decision-making, how their health and rights were ignored.[45] Between 2007 and 2010, they held meetings to share their stories and build a clearer picture of what they were facing.[46] These discussions helped the movement create stronger spaces for women's leadership.

That was how the *Arpilleras Project* was born. The idea came from a technique called *arpilleras*, which started in Chile in the 1970s, under the military dictatorship of Pinochet, when women began using fabric and thread to tell stories. They used pieces of cloth to create small scenes that

[45] Movimento Dos Atingidos Por Barragens. 2025. 'Mulheres e Arpilleras.' Available at: https://mab.org.br/en/women/ (Accessed: 20 June 2025).

[46] Ibid.

Fig. 2.2 Textiles produced by the *Arpilleras*. Source: MAB (2017, 2019). Movimento Dos Atingidos Por Barragens. 2017. 'Mulheres Atingidas Em Defesa Da Vida.' Available at: https://mab.org.br/arpilleras-do-mab/mulheres-atingidas-em-defesa-da-vida/ (Accessed: 29 June 2025); Movimento Dos Atingidos Por Barragens. 2019. 'Privatização Que Mata.' Available at: https://mab.org.br/arpilleras-do-mab/privatizacao-que-mata/ (Accessed: 29 June 2025).

showed what was happening in their communities: disappearances, violence and daily life under fear.[47] Some even included pieces of clothing from missing family members. These *arpilleras* were a way to speak out without words. Over time, *arpilleras* became a symbol of resistance and memory.

Inspired by this tradition, women in MAB began using *arpilleras* in Brazil to denounce the violence they experienced because of dams (Fig. 2.2). Since 2013, women have organised workshops in 19 states where they gather to sew together. Each piece tells a story: of a flooded home, a poisoned river, a broken community and a dream of a different future. The women choose the colours, the fabrics and the scenes. Each stitch is part of a bigger message: '*We are here. We will not be silent*'.[48]
But these workshops are more than just about making art. They are spaces for listening, healing and organising. Women share their pain but also their strength. They build connections and plan actions. The *arpilleras* help them tell their stories in their own way, especially in a society where women's voices are often ignored or seen as emotional and not political.[49]

[47] Ibid.

[48] Ibid.

[49] Rusansky, T. 2021. Embroidering Resistance: Emotional, Embodied, Everyday Struggles of Women Affected by Dams in Brazil. *Geoforum, 127*: 126–136. https://doi.org/10.1016/j.geoforum.2021.10.006

The *Arpilleras Project* shows that resistance does not always look like a march or a meeting. It can also happen around a table, with thread and cloth, with hands that carry both memory and hope. These women are not just telling their stories; they are changing them. As they say in MAB, *'Each stitch carries the belief that another world is necessary and possible'*.[50]

2.4.2 WoMin: A Pan-African Ecofeminist Alliance

Founded in October 2013, WoMin is a Pan-African ecofeminist alliance that works in solidarity with women and communities affected by extractivist-driven development and climate injustices. By combining 'women' and 'mining' , WoMin's name itself reflects its roots and core mission: to make visible the often-unseen impacts of extractive industries on women across the African continent.[51] It supports women's organising and movement-building and advances a vision for a Pan-African, post-extractivist, ecologically just and women-centred development model.[52]

Rooted in an ecofeminist perspective, WoMin recognises that the domination of nature and the oppression of women are fundamentally linked.[53] Ecofeminism, as both a political and intellectual tradition, critiques how patriarchal and capitalist systems treat both women and the environment as resources to be exploited.[54] It insists that women's rights and environmental justice must be pursued together, not at each other's expense. Ecofeminists emphasise that extractive and industrial development disproportionately affects women, while often excluding them from decision-making processes.[55]

[50] Movimento Dos Atingidos Por Barragens. 2025. 'Mulheres e Arpilleras.' Available at: https://mab.org.br/en/women/ (Accessed: 20 June 2025).

[51] WoMin. 2025. 'About Us.' Available at: https://womin.africa/about-us/ (Accessed: 15 June 2025); Hargreaves, S. *et al.* 2025. A Critical Exploration of Pan-African Ecofeminist Popular Education within WoMin's Feminist Schools. *Studies in the Education of Adults,* 1–20. https://doi.org/10.1080/02660830.2024.2445905

[52] WoMin. 2025. 'About Us.' Available at: https://womin.africa/about-us/ (Accessed: 15 June 2025)

[53] Mies, M. and Shiva, V. 1993. *Ecofeminism.* Zed Books.

[54] Buckingham, S. 2015. Ecofeminism. *International Encyclopedia of the Social & Behavioral Sciences.* Elsevier.

[55] Woehrle, L.M. 2022. Gender Studies. *Encyclopedia of Violence, Peace, & Conflict.* Elsevier.

Fig. 2.3 Advocacy materials from WoMin. Source: WoMin (2024). WoMin 2025. 'Women Building Power.' Available at: https://womin.africa/category/blog/women-building-power/. (Accessed: 15 June 2025).

WoMin operates across thirteen African countries, partnering with grassroots organisations in West, Central, East and Southern Africa.[56] Its emphasis on Pan-Africanism reflects a commitment to resisting colonial and neocolonial structures of power. The organisation sees Pan-Africanism not as an abstract political idea, but as a living and grounded process led by African women through collective struggle and solidarity.[57] In the energy sector, WoMin leads two major initiatives: *Women Building Power* and *Resisting Green Extractivism*. These programmes reveal how women across Africa are not only resisting energy-related injustices but also proposing alternatives.

The *Women Building Power* programme focuses on supporting resistance to mega energy projects such as fossil fuel plants and large dams. Women participating in this programme confront the destructive impacts of energy projects like the Sendou coal plant in Senegal, the Grand Inga

[56] Hargreaves, S. *et al.* 2025. A Critical Exploration of Pan-African Ecofeminist Popular Education within WoMin's Feminist Schools. *Studies in the Education of Adults,* 1–20. https://doi.org/10.1080/02660830.2024.2445905

[57] Ibid.

Fig. 2.4 WoMin's Right to Say No Programme. Source: WoMin (2024).

dam in the Democratic Republic of Congo, coal extraction in South Africa and gas infrastructure in Kenya.[58] These projects often displace communities, degrade local environments and deny women access to the energy they help to produce. At the same time, women's voices are excluded from decisions about their own energy futures.

In addition, WoMin also warns against the dangers of 'green' extractivism. In its *Resisting Green Extractivism* programme, the alliance draws attention to the ways that the global energy transition is replicating colonial patterns of resource extraction, as Africa is now seen as a key source of rare earths and critical minerals necessary for technologies such as solar panels, wind turbines and electric vehicles. Yet, these new green projects are often corporate-led and export-oriented and leave local communities behind.[59] In regions like Namakwaland in South Africa's Northern Cape, WoMin supports local communities resisting green hydrogen and mineral extraction projects that threaten to dispossess them under the banner of 'clean energy'. In this case, they supported the local community winning back communal land ownership and helped launch

[58] WoMin 2025. 'Women Building Power.' Available at: https://womin.africa/category/blog/women-building-power/ (Accessed: 15 June 2025).

[59] WoMin. 2025. 'Resisting Green Extractivism.' Available at: https://womin.africa/category/green-extractivism/ (Accessed: 27 June 2025).

the VVVT Namakwaland movement, which champions Free, Prior and Informed Consent.[60]

Therefore, WoMin does not stop at critique. It works with women to propose community-led energy solutions and reclaim control over land and decision-making processes, combining resistance with reconstruction, critique with creativity and local action with Pan-African solidarity. It highlights that energy and climate justice cannot be achieved without centring women's voices, respecting local sovereignty and addressing the deep legacies of extractivism and colonialism.

2.5 Discussion: Reclaiming Marginalised Territories, Bodies and Voices in Energy Justice

The *Arpilleras Project* and the *WoMin Pan-African Ecofeminist Alliance* represent two distinct yet complementary approaches to advancing energy justice through feminist and community-led approaches. While MAB operates within the national boundaries of Brazil, its reach spans a country of continental proportions, marked by deep regional diversity in culture, environment and political context. WoMin, in turn, works across multiple African countries, navigating different national governance systems and legal frameworks. Both initiatives mobilise women affected by large-scale energy and extractive projects, yet they differ in structure: MAB is a grassroots social movement organised from within communities directly impacted by dams, whereas WoMin is a cross-border alliance led by self-identified ecofeminists who coordinate regional campaigns and advocacy. Despite these differences, both organising share a commitment to amplifying the voices of women at the margins of dominant just energy transitions, centring lived experiences and proposing transformative, community-rooted alternatives to extractivist development.

The eight analytical dimensions described in the methodology help to illuminate how each initiative navigates and contests power, constructs knowledge and proposes alternatives from their own positionalities. Table 2.2 presents a comparative synthesis of how these dimensions manifest in the work of the MAB *Arpilleras Project* and WoMin.

This comparative analysis also serves as the basis for addressing this chapter's central research questions.

[60] Ibid.

Table 2.2 Case studies through the lens of the eight analytical dimension framework.

Analytical Dimension	*MAB Arpillera Project*	*WoMin*
Epistemic Justice	Reclaims space for women's embodied and situated knowledges through textile narratives that challenge technocratic and extractivist discourses.	Centres grassroots feminist and decolonial epistemologies, challenging the dominance of Northern, technocratic knowledge systems.
Intersectionality	Highlights the intersection of gender, race, class and territorial displacement in shaping women's experiences of energy injustice.	Engages with intersecting inequalities across gender, class, race and geography in diverse African contexts.
Structural Power	Critiques the centralised, profit-driven energy model through visual storytelling that exposes systemic inequality and exploitation.	Confronts global extractivist systems and colonial legacies by targeting structural sources of energy injustice.
Sovereignty	Fosters self-organised workshops where women reclaim narrative and political agency based on their own realities.	Supports women's rights to define development, reclaim land and assert collective energy and territorial sovereignty.
Intersecting Problems and Root Causes	Connects personal and community-level harms to systemic issues such as patriarchy, colonialism and capitalism.	Frames energy injustice within broader systemic violence, including dispossession, labour exploitation and ecological harm.
Transformative Solutions	Uses sewing workshops as spaces of healing and mobilisation, building feminist consciousness and proposing alternatives.	Invests in feminist education, convergence-building and community-led alternatives that reimagine African development.
Care and the Reproduction of Life	Centres care and collective healing as political acts, creating supportive spaces for emotional and social repair.	Embeds care and ecofeminist values in its organising, emphasising life-sustaining systems over capitalist growth.
Accountability	Transforms private pain into public testimony, demanding recognition and redress from institutions and society.	Holds states and corporations accountable through advocacy, legal strategies and regional campaigns for structural change.

Source (2025): Created by Arnal-Calvo, Beatriz; Freed Huici, Carmen; Schneider, Kathlen.

2.5.1 How Do Women-led Organising from the Global South Articulate Their Actions in Demand of Energy Justice?

Both MAB's *Arpilleras Project* and WoMin articulate energy justice as deeply connected to broader struggles for land, life and dignity. Their actions are not limited to calls for fairer energy access or infrastructure distribution. Instead, they centre resistance to extractivism and the reclaiming of territories, bodies and voices. The *Arpilleras Project* makes use of collective sewing workshops to narrate the gendered impacts of large dams, transforming memory into political action. WoMin's campaigns, in contrast, use advocacy, legal strategies, political education and cross-border alliances to confront both fossil fuel exploitation and the rising tide of 'green extractivism'. These actions reflect not only resistance but also **transformative solutions** that centre women's leadership, collective agency and knowledge rooted in lived experience.

2.5.2 How Do Their Lived Experiences, Embodied Knowledges and Situated Demands Inform Those Actions?

Here, the dimensions of **epistemic justice**, **intersectionality** and **care and the reproduction of life** are central. Both organising foreground the everyday experiences of rural, Indigenous, Black and working-class women, whose stories are often excluded from formal energy policy and academic discourse. MAB women narrate their realities through *arpilleras*, challenging the masculine, technocratic language of energy planning with emotional, embodied and situated forms of knowledge. WoMin similarly elevates women's experiences of environmental destruction and social violence through feminist political education and participatory research. In both cases, these lived realities not only inform their critique of existing systems but also guide alternative visions of well-being, care and sustainability – where energy systems are designed to support life rather than extract from it.

2.5.3 How Do These Organising Contribute, If At All, to Existing but Narrow Hegemonic Energy Justice Frameworks – by Reproducing, Contesting, Resisting, Amplifying or Complementing Them?

Both case studies clearly **contest and expand** narrow energy justice frameworks, particularly those that focus primarily on distribution or procedural inclusion without addressing **structural power**. MAB's work, for example, brings emotion, memory and art into political space, unsettling traditional boundaries between the personal and the political. WoMin contests the notion that renewable energy is inherently just, exposing how 'green' projects can replicate extractive logics if led by corporate and state elites. In doing so, both initiatives amplify alternative frameworks that emphasise **sovereignty**, self-determination and collective care. They propose an expanded vision of energy justice that includes epistemic, ecological and gendered dimensions.

2.5.4 How Can Their Contributions Advance More Inclusive and Transformative Energy Justice Research and Policies?

The contributions of these organising offer concrete entry points for more inclusive and transformative approaches to energy justice research and policymaking. Drawing from the dimensions of **transformative solutions**, **accountability** and **intersecting root causes**, both MAB and WoMin show that justice must be structural, not symbolic. First, they highlight the need to engage with justice not only as an outcome but as a process rooted in historical context, collective memory and everyday struggle. Second, they show that policies must be informed by those who have traditionally been excluded from decision-making spaces, particularly rural, Indigenous, Black and working-class women. Third, they call for a fundamental shift in how knowledge is valued, recognising that expertise does not only reside in institutions but also in communities, traditions and embodied practices.

2.6 Final Remarks: Towards Inclusive and Transformative Energy Justice Frameworks

This chapter showed that incorporating decolonial, intersectional and feminist perspectives attentive to epistemic justice is a way of advancing towards more inclusive and transformative just energy frameworks. By interrogating who the political subjects and knowledge producers of energy justice are, it amplified contributions from two women-led, grassroots organising in the Global South rooted in their lived experiences, embodied knowledges and situated demands, therefore rethinking justice. Both the *Arpilleras Project* in Brazil and WoMin in the African continent challenge the historical systems of oppression embedded in energy systems, opening space for justice to be reimagined otherwise: from the ground up. Further, they acknowledge fundamental contributions from experiences that are embodied, lived and situated, challenging epistemological hegemonies and building pluralistic understandings of energy and justice. Consequently, they diversify the range of knowledge and perspectives involved in energy decision-making.

Building on this novel energy justice framework and two case studies, this chapter foregrounds the following policy and research recommendations: (i) employ decolonial, intersectional and feminist analyses in all energy justice processes, recognising the diversity of women's identities and roles along with the structural power relations that shape their (lack of) participation and outcomes; (ii) recognise women and specially indigenous women and women of colour as energy justice agents and knowledge producers by centring the lived experiences, embodied knowledges and situated demands of women-led grass roots organising in the Global South; (iii) include them in the design, implementation and monitoring of energy transitions programmes; (iv) prioritise direct funding and capacity building for women-led, grass roots organising in the Global South; (v) recognise and favour feminist, local, decolonial, Indigenous, community-based and critical perspectives to energy justice over hegemonic technocratic and top-down approaches; (vi) address and redress the structural (e.g. legal, financial and cultural) obstacles to women's access to and participation in energy systems; (vii) instead of tokenistic women's participation, assess and promote transformative outcomes such as shifts in power relations, increased agency and the realisation of women's collective rights and (viii) implement grass roots-led participatory evaluation methods that ensure accountability and life-sustaining praxes. Ultimately, energy justice policy and research ought to lay the groundwork for just knowledge, energy and power transformations.

CHAPTER 3

Democratising the EU Nuclear Energy Policy Through Energy Justice

Zamira Xhaferri, Filiz Doğan, Emiliano Castillo Jara, Gianna Giardini, and Hazal Mengi

Abstract Russia's invasion of Ukraine in 2022 and the following energy crisis have reshaped the EU's energy strategy. Confronted with the challenge of achieving carbon neutrality while ensuring energy security, the nuclear energy narrative has prevailed in academic and policy debates. While some EU member states consider it essential to the EU's energy

Z. Xhaferri
European Law, European Studies Department, Universiteit van Amsterdam, Amsterdam, The Netherlands
e-mail: z.xhaferri@uva.nl

F. Doğan
Faculty of Governance & Global Affairs, Universiteit Leiden, Leiden, The Netherlands

E. Castillo Jara
Department of Human Geography and Environmental Sciences, Universität Trier, Trier, Germany

G. Giardini
Istituto Universitario di Studi Superiori Università di Pavia and Università di Camerino, Pavio, Italy

H. Mengi
Department of European Union, Istanbul University, Istanbul, Türkiye

A. Phillips, L. Kaschny (eds.), *Global Perspectives on the Circular Economy in the Just Transition*, Just Transitions,
https://doi.org/10.1007/978-3-032-18342-2_3

future, others advocate for its phase-out, limiting the EU's ability to streamline its energy policy. While Small Modular Reactors (SMRs) offer flexibility and shorter deployment times, they raise concerns about waste, safety and regulatory consistency. In that vein, legal, technical, economic and geopolitical concerns prevail, while public participation is often overlooked to the detriment of democracy and accountability. Our research examines the EU's nuclear governance framework through the lens of energy justice. It investigates who benefits from nuclear development and who bears its risks, whether the needs of future generations are considered in decision-making and to what extent current nuclear energy policies address structural inequalities among EU member states. Our research findings suggest rethinking a future-proof regulatory framework that integrates legal clarity, technological innovation and democratic legitimacy to ensure a just energy transition in the European Union.

Keywords Nuclear Energy • Energy Law • Energy Governance • Energy Communities • Energy Justice • Democratic Legitimacy • Energy Transition

3.1 Introduction

Russia's invasion of Ukraine in 2022 has exposed the European Union's (EU) energy security vulnerabilities leading to increased calls to reduce its dependence on Russian fossil fuels and reconsider its energy strategy. Under the Fit for 55 packages, the EU aims to reduce its greenhouse gas emissions (GHG)[1] by 55% by 2030, achieve climate neutrality by 2050,[2] and raise the share of renewable energy to at least 45% by 2030. In 2020, the EU adopted the Taxonomy Regulation,[3] which classifies nuclear energy

[1] For a discussion on the global importance of reducing GHG see Skukla, P. R. *et al.* (eds) In: *Climate Change 2022: Mitigation of Climate Change. Contribution of Working Group III to the Sixth Assessment Report of the Intergovernmental Panel on Climate Change*, ed. P. R. Shukla *et al.*, 1–42. Cambridge: Cambridge University Press. https://doi.org/10.1017/9781009157926.001

[2] Communication from the Commission to the European Parliament, the European Council, the Council, the European Economic and Social Committee and the Committee of the Regions. REPowerEU Plan (COM/2022/230 final), Brussels.

[3] Regulation (EU) 2020/852 of the European Parliament and of the Council of 18 June 2020 on the establishment of a framework to facilitate sustainable investment and amending Regulation (EU) 2019/2088 (Taxonomy Regulation). OJL 198, 22.6.2020.

as 'sustainable' under certain conditions. As renewables are positioned as a central pillar of the EU's energy policy, these targets place additional pressure on policymakers to balance energy security and decarbonisation goals. These events have triggered debates about the role of nuclear power as a key source for the EU energy policy.[4] Nuclear power has long been criticised due to several concerns, including risks of accidents, long-term waste disposal, water pollution, health risks and high economic costs.[5] Driven by the economic growth imperative, the nuclear sector often externalises risks and costs to the public, while concentrating profits among private and state companies.[6] Marginalised communities have disproportionately borne the environmental and social risks associated with nuclear infrastructure, thereby creating structural inequalities.[7]

The EU member states remain divided. Pro-nuclear states view nuclear energy as essential for long-term energy security, independence and decarbonisation. Others seek to phase it out to benefit renewables.[8] The inclusion of nuclear power as a sustainable economic activity within the EU climate and

[4] Article 194(2) TFEU; Heiko, K. 2016. European Energy Law and Policy: An Introduction. Cheltenham, UK: Edward Elgar Publishing; Huhta, K. 2025. An Essential Zero-Carbon Solution or a Perilous Energy Technology? Exploring the nuclear debate in EU law. In: Reins, L. and Zahar, A. (eds) *Climate Technology and Law in the Anthropocene*. Bristol: Bristol University Press. See also Heffron, R. J. 2015. *Deconstructing Energy Law and Policy*. Edinburgh: Edinburgh University Press.

[5] Peterson, M. 2024. *Derrida and Inheritance in Environmental Ethics: The half-lives of responsibility*. Cham: Palgrave Macmillan. https://doi.org/10.1007/978-3-031-52143-0; Höffken, J. and Ramana, M.V. 2024. Nuclear Power and Environmental Injustice. *Wiley Interdisciplinary Reviews: Energy and Environment, 13*(1): https://doi.org/10.1002/wene.498

[6] Heffron, R. 2015. *Deconstructing Energy Law and Policy: The case of nuclear energy*. Edinburgh: Edinburgh University Press; Cotton, M. 2021. Nuclear Power and Environmental Justice: The case for political equality. In: Kaiser, A., Lehtonen, M., Meyer, J. H. & Rubio, M. (eds). *Engaging the Atom: The history of nuclear energy and society in Europe from the 1950s to the present*, Morgantown: West Virginia University Press.

[7] Höffken, J. and Ramana, M. V. 2024. Nuclear Power and Environmental Injustice. *Wiley Interdisciplinary Reviews: Energy and Environment, 13*(1). https://doi.org/10.1002/wene.498

[8] Austria, Germany, Spain, and Nordic and Iberian countries; Todorović, I. 2024. EU remains divided over nuclear energy expansion initiative. *Balkan Green Energy News*. Available at: https://balkangreenenergynews.com/eu-remains-divided-over-nuclear-energy-expansion-initiative/ (Accessed: 28 May 2025); Judah, B. *et al.* 2023. How to Strike a Grand Bargain on EU Nuclear Energy Policy. *Atlantic Council*. Available at: https://www.atlanticcouncil.org/in-depth-research-reports/report/how-to-strike-a-grand-bargain-on-eu-nuclear-energy-policy/. (Accessed: 15 May 2025).

frameworks indicates a policy shift.[9] Simultaneously, the EU is promoting the development of citizen-driven energy models encompassing energy communities to ensure citizens' engagement and an equal distribution of risks and benefits in energy deployment. This raises the question of how nuclear energy could be developed within a decentralised market.[10] Worthy of note are the Small Modular Reactors (SMRs),[11] a fourth-generation nuclear technology that has emerged alongside the growing development of decentralised energy production and consumption models.[12]

[9] Regulation (EU) 2020/852 and Commission Delegated Regulation (EU) 2022/1214 have exacerbated debates over its environmental risks and potential to reduce GHG emissions; Bohdanowicz, Z., Łopaciuk-Gonczaryk, B., and Kowalski, J. 2022. Europe Becomes Pro-nuclear? Drivers of public support for nuclear energy in six EU countries after the energy crisis of 2022. 2025. *Energy Policy, 200*: https://doi.org/10.1016/j.enpol.2025.114528

[10] The Clean Energy for All Europeans Package, adopted between 2018 and 2019, updated the EU energy law comprising eight legislative acts, namely: (i) Directive (EU) 2018/844 of the European Parliament and of the Council of 30 May 2018 amending Directive 2010/31/EU on the energy performance of buildings and Directive 2012/27/EU on energy efficiency, OJ L 156, 19.6.2018; (ii) Directive (EU) 2018/2001 of the European Parliament and of the Council of 11 December 2018 on the promotion of the use of energy from renewable sources (recast) (RED II), as amended until 2024, OJ L 328, 21.12.2018; (iii) Directive (EU) 2018/2002 of the European Parliament and of the Council of 11 December 2018 amending Directive 2012/27/EU on energy efficiency, OJ L 328, 21.12.2018; (iv) Directive (EU) 2019/944 of the European Parliament and of the Council of 5 June 2019 on common rules for the internal market for electricity and amending Directive 2012/27/EU, as amended until 2024, OJ L 158, 14.6.2019; (v) Regulation (EU) 2018/1999 of the European Parliament and of the Council of 11 December 2018 on the Governance of the Energy Union and Climate Action, OJ L 328, 21.12.2018; (vi) Regulation (EU) 2019/943 of the European Parliament and of the Council of 5 June 2019 on the internal market for electricity, OJ L 158, 14.6.2019; (vii) Regulation (EU) 2019/941 of the European Parliament and of the Council of 5 June 2019 on risk-preparedness in the electricity sector and repealing Directive 2005/89/EC, OJ L 158, 14.6.2019; (viii) 1 Regulation (EU) 2019/942 of the European Parliament and of the Council of 5 June 2019 establishing a European Union Agency for the Cooperation of Energy Regulators, OJ L 158, 14.6.2019.

[11] Nuclear reactors are generally categorised by generations, namely, I, II, II and IV. SMRs are included into the IV generation. Earlier nuclear generations include: Generation I (1950s-1960s) prototype and demonstration reactors, generation II (1970s-1980s) large light and heavy water reactors, and generation III (1990s-present) advanced reactors with developed safety systems and efficiency. See Goldberg, S. M., & Rosner, R. 2011. Nuclear Reactors: Generation to Generation. *American Academy of Arts and Sciences*. Available at: https://www.amacad.org/sites/default/files/academy/pdfs/nuclearReactors.pdf. (Accessed: 11 August 2025); World Nuclear Association. 2024. Advanced Nuclear Power Reactors. London: World Nuclear Association. Available at: https://world-nuclear.org/information-library/current-and-future-generation/outline-history-of-nuclear-energy (Accessed: 11 August 2025).

[12] European Commission. 2016. Clean Energy for All Europeans (COM (2016) 860 final). Available at: https://eur-lex.europa.eu/legal-content/EN/TXT/HTML/?uri=CELEX:52016DC0860 (Accessed: 20 October 2025).

Given these challenges, a comprehensive reassessment of the EU's nuclear policy must go beyond its technical and geopolitical considerations to encompass normative ones. Integrating principles into nuclear policy, such as equity, democratic participation and intergenerational responsibility, could contribute to reducing social inequalities and aligning the EU's aims with energy justice. The EU's nuclear policy has so far been under-researched from an energy perspective.[13]

Our chapter examines how the principles of energy justice encompassing distributive, procedural, recognition, restorative and cosmopolitan, unveil and address the multifaceted inequalities embedded in the EU's nuclear policy. It critically engages with academic literature, nuclear policy documents and the legal framework governing nuclear energy. Applying the principles of energy justice allows for a critical assessment of whose interests and perspectives are represented in decision-making on nuclear energy and how the benefits and burdens of nuclear power are distributed. Section 2 outlines the analytical framework governing the EU's nuclear energy policy. Section 3 explores legal and policy tensions. It zooms in on the challenges posed by new energy technologies encompassing the energy communities and SMRs. Section 4 analyses their benefits and limitations through the lens of energy justice. Section 5 discusses the relevance of energy justice with a view to enhancing public participation and addressing inequalities in the EU's nuclear governance framework.

3.2 Conceptual Framework: The EU Nuclear Policy Through the Lens of Energy Justice

Nuclear energy has returned to the centre of EU energy debates with the 'Declaration to Triple Nuclear Energy Capacity by 2050' signed at the United Nations Climate Change Conference in Dubai (2023).[14] Analysing the EU's reconsideration of its nuclear power policy using the framework of energy justice sheds light on the role of power relations in (re)producing various forms of inequality and injustice associated with nuclear development in the EU. Energy justice builds upon five tenets:

[13] Kaschny, L. 2023. Energy Justice and the Principles of Article 194(1) TFEU Governing EU Energy Policy. *Transnational Environmental Law*: *23*(2): 270–294. doi:10.1017/S2047102523000110; Longeran, K. *et al.* 2022. Energy Justice Analysis of the European Commission REPowerEU plan. *Working Paper*. Zurich: ETH Zurich. https://doi.org/10.3929/ethz-b-000551952

[14] See also: European Parliament resolution of 21 November 2023 on the UN Climate Change Conference 2023 in Dubai, United Arab Emirates (COP28) (2023/263).

distributive, procedural, recognition, restorative and cosmopolitan justice.[15] Distributive justice addresses the unequal distribution of the benefits, costs and risks of energy development.[16] In the case of nuclear power, energy justice literature has documented the disproportionate risks posed by uranium mining, waste management and nuclear decommissioning to vulnerable communities, ecosystems and future generations.[17]

Procedural justice entails the design and implementation of fair and effective legal and policy mechanisms to ensure representation and participation in decision-making in the energy sector.[18] Nuclear power entails the provision of equal access to information for nuclear energy development and access to justice in energy policy-related disputes and decisions.[19] For instance, grassroots environmental movements in Germany, Denmark and Austria[20] have called for enhanced transparency and citizens' participation in decision-making for nuclear energy. They have also proposed a phase-out of nuclear energy due to safety, environmental and economic concerns.[21] Following these past controversies, energy justice draws attention to the role of social perceptions, values and identities in shaping public attitudes towards nuclear power.[22] Recognition justice also aims to ensure

[15] Heffron, R. J. and McCauley, D. 2017. The Concept of Energy Justice Across the Disciplines. *Energy Policy* 105*:* 658–667. https://doi.org/10.1016/j.enpol.2017.03.018

[16] Healy, N. *et al.* 2019. Embodied Energy Injustices: Unveiling and politicizing the transboundary harms of fossil fuel extractivism and fossil fuel supply chains. *Energy Research & Social Science, 48*: 219–234. https://doi.org/10.1016/j.erss.2018.09.016

[17] Heffron, R. 2015. *Deconstructing Energy Law and Policy: The case of nuclear energy.* Edinburgh: Edinburgh University Press.

[18] Heffron, R. J., and McCauley, D. 2014. Achieving Sustainable Supply Chains Through Energy Justice. *Applied Energy, 123:* 435–437. https://doi.org/10.1016/j.apenergy.2013.12.034

[19] Cotton, M. 2021. Nuclear Power and Environmental Justice: The case for political equality. In: Kaiser, A., Lehtonen, M., Meyer, J. H. & Rubio, M. (eds). *Engaging the Atom: The History of Nuclear Energy and Society in Europe from the 1950s to the Present.* Morgantown: West Virginia University Press.

[20] Puig, A. P. and Meyer, J. H. 2021. One Movement or Many? The Diversity of Antinuclear Movements in Europe. In: Kaiser, A., Lehtonen, M., Meyer, J. H. & Rubio, M. (eds). *Engaging the Atom: The History of Nuclear Energy and Society in Europe from the 1950s to the Present.* Morgantown: West Virginia University Press.

[21] Heffron, R. J., and McCauley, D. 2014. Achieving sustainable supply chains through energy justice. *Applied Energy, 123*, 435–437. https://doi.org/10.1016/j.apenergy.2013.12.034

[22] Bohdanowicz, Z. *et al.* 2023. Support for Nuclear Power and Pro-environmental Attitudes: The cases of Germany and Poland. *Energy Policy, 177.* https://doi.org/10.1016/j.enpol.2023.113578; Brook, B. *et al.* 2014. Why Nuclear Energy is Sustainable and Has to

that the rights, interests, cultural identities and perspectives of marginalised groups are represented in energy policies and law.[23] This means, for example, considering the needs, rights and perceptions of communities living near nuclear facilities and future generations throughout the planning, implementation and assessment of nuclear energy projects.[24<?><?><?>]

Acknowledging the historical and ongoing inequalities of energy development, restorative justice aims to prevent and mitigate their causes and compensate those affected.[25] Thus, governments and corporations would be held accountable for the long-lasting damages caused by uranium extraction, nuclear facilities, radioactive waste management and accidents following the use of nuclear energy.[26]

Given power imbalances in energy governance, cosmopolitan justice posits that, given the global character of climate change and its impacts on all citizens of the world, all nation-states have the obligation to meet their climate targets under the 2015 Paris COP21 Agreement to tackle this issue. Furthermore, cosmopolitan justice draws attention to the requirement to safeguard the rights of the next generations for environmental protection.[27] Applying this tenet to the nuclear power framework draws attention not only to the need for equal access to energy but also to the moral and legal obligations of states to prevent and restrict nuclear weapons proliferation.[28] Overall, energy justice highlights the inadequacy of legal-institutional arrangements to address a wide range of concerns

be part of the Energy Mix. *Sustainable Materials and Technologies,* 1–2: 8–16. https://doi.org/10.1016/j.susmat.2014.11.001

[23] Heffron, R. J., and McCauley, D. 2014. Achieving sustainable supply chains through energy justice. *Applied Energy*, *123*, 435–437.https://doi.org/10.1016/j.apenergy.2013.12.034

[24] Peterson, M. 2024. *Derrida and Inheritance in Environmental Ethics: The half-lives of responsibility*. Cham: Palgrave Macmillan. https://doi.org/10.1007/978-3-031-52143-0

[25] Sokołowski, M. M. and Heffron, R. J. 2022. Defining and Conceptualising Energy Policy Failure: The when, where, why, and how. *Energy Policy*, *161*. https://doi.org/10.1016/j.enpol.2021.112745; Hazrati, M. and Heffron, R. J. 2021. Conceptualising Restorative Justice in the Energy Transition: Changing the perspectives of fossil fuels. *Energy Research & Social Science,* *78*: https://doi.org/10.1016/j.erss.2021.102115; Heffron, R. J., and Hazrati, M. 2024. *The Rise of Energy Justice Research in the Energy Transition and for Climate Mitigation*. Just Transitions. Palgrave Macmillan, Cham. https://doi.org/10.1007/978-3-031-57304-0_1

[26] Peterson, M. 2024. *Derrida and Inheritance in Environmental Ethics: The half-lives of responsibility*. Cham: Palgrave Macmillan. https://doi.org/10.1007/978-3-031-52143-0

[27] Heffron, R. J., and De Fontenelle, L. 2023. Implementing Energy Justice Through a New Social Contract. *Journal of Energy & Natural Resources Law*, *41*(2): 141–155. https://doi.org/10.1080/02646811.2023.2186626

[28] Heffron, R. J. and Sokolowski, M. M. 2024. Resolving Energy Policy Failure: Introducing energy justice as the solution to achieve a just transition. *Energy Policy*, *187*. https://

and inequalities associated with nuclear energy development. Adopting the energy justice perspective points to the need to redefine EU nuclear energy governance based on justice considerations. This approach moves beyond dominant legal-institutional analysis focusing on technical, economic, and geopolitical aspects of nuclear policy. The inclusion of energy justice in nuclear policies, however, may lead to tensions and contradictions between competing interpretations of justice.[29] For example, forms of public participation in nuclear policy may legitimise decisions already taken by powerful state and corporate actors. Considering these contradictions is essential to critically evaluate nuclear governance in the EU.

3.2.1 Regulating Emergent Nuclear Power Approaches and Technologies: Energy Communities and Small Modular Reactors

Nuclear energy is an important low-carbon source of electricity in the EU, as it generates low greenhouse gas emissions during operation. The Taxonomy Regulation (2020)[30] frames nuclear energy as 'sustainable' activity under strict conditions that were further detailed in a number of non-legislative acts adopted by the European Commission.[31] [32] In its

doi.org/10.1016/j.enpol.2024.114042; Heffron, R. 2015. *Deconstructing Energy Law and Policy: The case of nuclear energy*. Edinburgh: Edinburgh University Press.

[29] McHarg, A. 2020. Energy Justice: Understanding the 'ethical turn' in energy law and policy. In: del Guayo, I. *et al.* (eds). *Energy Justice and Energy Law*. Oxford: Oxford University Press; Heffron, R. J. *et al.* 2023. Pathways of Scholarship for Energy Justice and the Social Contract. *Journal of Energy & Natural Resources Law*: *41*(2): 211–232. https://doi.org/10.1080/02646811.2023.2190689

[30] Regulation (EU) 2020/852 of the European Parliament and of the Council of 18 June 2020 on the establishment of a framework to facilitate sustainable investment, and amending Regulation (EU) 2019/2088, OJL, 198, 22.6.2020.

[31] Commission Delegated Regulation (EU) 2021/2139 of 4 June 2021, OJ L 442, 9.12.2021; Commission Delegated Regulation (EU) 2021/2178 of 6 July 2021, OJ L 443, 10.12.2021; Commission Delegated Regulation (EU) 2022/1214 of 9 March 2022 (covering specific gas and nuclear activities), OJ L 188, 15.7.2022; Commission Delegated Regulation (EU) 2023/2486 of 27 June 2023, plus related amendment act (EU) 2023/2485, OJ L, 2023/2486, 21.11.2023,A Commission Delegated Act amending the Disclosures, Climate and Environmental Delegated Acts adopted 4 July 2025. The changes are set to apply from January 1, 2026, with the option to apply them to the 2025 financial year.

[32] Commission Delegated Regulation (EU) 2021/2139 of 4 June 2021, OJ L 442, 9.12.2021; Commission Delegated Regulation (EU) 2021/2178 of 6 July 2021, OJ L 443, 10.12.2021; Commission Delegated Regulation (EU) 2022/1214 of 9 March 2022

Action Plan for Affordable Energy,[33] the European Commission committed to streamlining by 2026 the procedures for permitting advanced nuclear technologies, including SMRs. This initiative reflects the EU's attempt to reconcile nuclear developments with its broader decarbonisation agenda. The governance of nuclear energy in the EU, however, continues to rest on the Euratom Treaty (1957)[34] followed by specific legislation on nuclear safety,[35] waste management,[36] and radiation protection.[37] Despite the evolving technological and regulatory dynamics of the European electricity market as set out by the legislative package 'Clean Energy for All Europeans' (2019),[38] there were tensions between the ambition to integrate nuclear governance into the clean energy transition and the limitations of the existing legal framework. The Renewable

(covering specific gas and nuclear activities), OJ L 188, 15.7.2022; Commission Delegated Regulation (EU) 2023/2486 of 27 June 2023, plus related amendment act (EU) 2023/2485, OJ L, 2023/2486, 21.11.2023,A Commission Delegated Act amending the Disclosures, Climate and Environmental Delegated Acts adopted 4 July 2025. The changes are set to apply from January 1, 2026, with the option to apply them to the 2025 financial year.

[33] Communication from the Commission to the European Parliament, the Council, the European Economic and Social Committee and the Committee of the Regions. Action Plan for Affordable Energy Unlocking the true value of our Energy Union to secure affordable, efficient and clean energy for all Europeans, COM(2025) 79 final, Brussels. On this point, see Abousahl, S. *et al.* 2021. *Technical assessment of nuclear energy with respect to the 'do no significant harm' criteria of Regulation (EU) 2020/852 ('Taxonomy Regulation')*. Publications Office of the European Union; Handrlica, J., and Blahoudková, G. 2025. Microreactors for powering data centres in smart cities, miniaturisation of technologies and the paradigm shift in nuclear law. Available at: https://ceridap.eu/microreactors-for-powering-data-centres-in-smart-cities-miniaturisation-of-technologies-and-the-paradigm-shift-in-nuclear-law/ (Accessed: 29 June 2025).

[34] The Treaty establishing the European Atomic Energy Community, Rome, 25 March 1957.

[35] Council Directive 2009/71/Euratom of 25 June 2009 establishing a Community framework for the nuclear safety of nuclear installations, OJL 172, 2.7.2009.

[36] Council Directive 2011/70/Euratom of 19 July 2011 establishing a Community framework for the responsible and safe management of spent nuclear fuel and radioactive waste, OJ L 199, 2.8.2011.

[37] Council Directive 2013/59/Euratom of 5 December 2013, laying down basic safety standards for protection against the dangers arising from exposure to ionising radiation, OJL 13, 17.1.2014.

[38] Edwards, R. 2004. Nuclear disarray as Europe pushes east. New Scientist. Available at: https://www.newscientist.com/article/mg18224451-000-nuclear-disarray-as-europe-pushes-east/ (Accessed: 11 August 2025).

Energy and Electricity Market Directives, along with their amendments[39] established a decentralised, inclusive and participatory model in which EU citizens play an active role.[40] Advances in accessible technologies have fundamentally transformed the EU's energy landscape, opening it up to distributed and participatory systems. What it was once a field dominated by large and centralised utilities it has evolved into a space where citizens, small enterprises and local authorities collaborate towards cooperative models of energy generation and sharing. This paradigm shift raises important questions about the compatibility of nuclear power, which remains inherently centralised, with a more democratic and locally driven energy governance framework.

Energy communities as legal entities embody a new model of governance following principles of voluntary participation, openness and effective control by local bodies. They comprise a wide range of activities consisting of energy production and supply, energy sharing, local grid management, promotion of energy efficiency, energy storage and the provision of ancillary services to create environmental, economic, or social benefits for community members and their territories.[41] The energy is used for self-consumption and shared among community members. Any surplus energy can be sold and the income derived from the energy communities could be reinvested in local projects for economic, social and environmental benefits.[42] Energy communities contribute to social and economic

[39] Directive (EU) 2018/2001 of the European Parliament and of the Council of 11 December 2018 on the promotion of the use of energy from renewable sources (recast) (RED II), as amended until 2024, OJL 328, 21.12.2018. See also: Directive (EU) 2024/1711 of the European Parliament and of the Council of 13 June 2024 amending Directives (EU) 2018/2001 and (EU) 2019/944 as regards improving the EU's electricity market design, OJ L, 2024/1711, 26.6.2024. Directive (EU) 2019/944 of the European Parliament and of the Council of 5 June 2019 on common rules for the internal market for electricity and amending Directive 2012/27/EU (as amended until 2024), OJL 158, 14.6.2019. See also: Directive (EU) 2024/1711 of the European Parliament and of the Council of 13 June 2024 amending Directives (EU) 2018/2001 and (EU) 2019/944 as regards improving the EU's electricity market design, OJ L, 2024/1711, 26.6.2024.

[40] Ruggeri, L. 2021. Legislative Policies and Jurisprudence on Climate Change: New Tools for Removing the Barriers to New Forms of Energy Consumption. In: Ruggeri, L. (eds), 2021. Needs and Barriers of Prosumerism in the Energy Transition Era, Dickinson, 98–109.

[41] Article 2 (16) in Directive (EU) 2018/2001 read together with Article 2 (11), Directive (EU) 2019/944.

[42] Gui, E. M. and MacGill, I. 2018. Typology of future clean energy communities: An exploratory structure, opportunities, and challenges. *Energy Research & Social Science, 35*: 94–107. https://doi.org/10.1016/j.erss.2017.10.019

growth through a new governance model involving decision-making engaged consumers to mitigate the power asymmetries between energy producers and consumers.[43]

Unlike the Renewable Energy Communities (RECs),[44] which focus exclusively on renewable sources, the Citizens' Energy Communities (CECs) can draw on renewable and non-renewable energy sources according to the principle of democratic participation of community members.[45] This delineation is important to ensure an inclusive energy transition, one that does not exclude, *a priori*, the deployment of nuclear energy, provided it operates within a solid regulatory framework aligned with sustainability and safety goals.

From this perspective, energy communities could be an innovative mechanism for 'democratising' nuclear energy by promoting a model based on participation and local engagement, thereby increasing social acceptance of a technology historically perceived as dangerous. The aim would be not only to decentralise production but also to redefine its scope and methods according to principles of sharing, care and collective responsibility. This shift places local communities at the centre of decision-making, thereby decreasing historical shortcomings, such as vertical governance structures and the absence of meaningful dialogue with affected communities.

Most energy communities focus on renewable energy. However, recent advances in nuclear technology, such as SMRs, offer new possibilities. Although technological developments linked to SMRs date back to the 1950s, interest in their development has grown since the 2011 Fukushima nuclear accident.[46] SMRs are advanced nuclear reactors with a power capacity of up to 300 MW(e) per unit, which is about one-third of the 1,000

[43] Hanke, F., and Guyet, R. 2023. The Struggle of Energy Communities to Enhance Energy Justice: Insights from 113 German cases. *Energy Sustainable Society, 13*, 16. https://doi.org/10.1186/s13705-023-00388-2; Olawuyi, D. From Energy Consumers to Energy Citizens Legal Dimensions of Energy Citizenship. In: Fleming, R., Huhta K., and Reins, L., (eds). *Sustainable Energy Democracy and the Law*, Brill, 101–123. https://doi.org/10.1163/9789004465442_006

[44] Article 22 in Directive (EU) 2018/2001.

[45] Article 16 in Directive (EU) 2019/944; Caramizaru, A. and Uihlein, A., 2020. Energy communities: an overview of energy and social innovation. Publications Office of the European Union, Luxembourg.

[46] Sovacool, B. K., and Ramana, M.V. 2015. Back to the Future: Small Modular Reactors, Nuclear Fantasies, and Symbolic Convergence. *Science, Technology, & Human Values*: *40*(1): 96–125. http://www.jstor.org/stable/43671221

MW generating capacity of traditional nuclear power reactors.[47] They are promoted as promising low-carbon energy solutions aligned with the Paris Agreement,[48] with proponents suggesting they could help countries achieve their climate targets and address key challenges in nuclear power, such as high upfront construction costs, potential accidents, water pollution and radioactive waste generation.[49] For example, due to their flexible siting requirements, SMRs can be deployed in off-grid or weak-grid regions, reducing transmission costs and strengthening nuclear facilities' resilience to disruptions.[50] In addition, the International Atomic Energy Agency (IAEA) underscores the role of SMRs in decarbonising energy-intensive sectors through both electricity generation and heating.[51]

While SMRs offer advantages over large reactors (LRs), they also face notable limitations.[52] For instance, the cost competitiveness of SMRs compared to LRs is still a concern. Subsidies are therefore essential for developing the SMR market and reducing financial uncertainty.[53] Additionally,

[47] International Atomic Energy Agency (IAEA). 2023. What are Small Modular Reactors (SMRs)? Vienna: IAEA. Available at: https://www.iaea.org/newscenter/news/what-are-small-modular-reactors-smrs (Accessed: 6 May 2025); SMRs have smaller compact sizes requiring less space and cooling water and shorter construction timelines of 3–4 years compared to 10 years for traditional NPPs with lower up-front capital costs; Ernst and Young (EY). 2024. The True Power of Small Modular Reactors on the Road to A Sustainable Energy Future: Unveiling key opportunities and challenges Paris: EY. Available at: https://www.ey.com/content/dam/ey-unified-site/ey-com/fr-fr/insights/energy-resources/documents/ey-etude-nuclear-energy-smr-pov-20240322.pdf (Accessed: 22 May 2025).

[48] The United Nations Framework Convention on Climate Change. Paris Agreement of 12 December 2015.

[49] IAEA. 2022. Small Modular Reactors: A new nuclear energy paradigm. Vienna: IAEA. Available at: http://nucleus.iaea.org/sites/smr/Shared%20Documents/Small%20Modular%20Reactors%20a%20new%20nuclear%20energy%20paradigm.pdf (Accessed: 8 May 2025).

[50] Gaster, R. 2025. Small Modular Reactors: A Realist Approach to the Future of Nuclear Power. *Center for Clean Energy Innovation, Information Technology and Innovation Foundation (ITIF)*. Washington DC: ITIF. Available at: https://itif.org/publications/2025/04/14/small-modular-reactors-a-realist-approach-to-the-future-of-nuclear-power/ (Accessed: 8 August 2025).

[51] IAEA. 2021. Nuclear Energy for a Net-Zero World. Vienna: IAEA. Available at: https://www.iaea.org/sites/default/files/21/10/nuclear-energy-for-a-net-zero-world.pdf (Accessed: 11 May 2025).

[52] European Parliament. 2023. Small Modular Reactors: Opportunities and Challenges for the EU, Policy Department for Industry, Research and Energy. Available at: https://energy.ec.europa.eu/topics/nuclear-energy/small-modular-reactors_en (Accessed:12 May 2025).

[53] IAEA. 2021. Nuclear Energy for a Net-Zero World. Vienna: IAEA. Available at: https://www.iaea.org/sites/default/files/21/10/nuclear-energy-for-a-net-zero-world.pdf (Accessed: 11 May 2025).

emerging harmonisation and standardisation of SMR licensing processes are often perceived as regulatory barriers.[54] Furthermore, although the EU supports SMR-related research[55]<?> through diverse mechanisms, such as Euratom and Horizon Europe, some initiatives like the European Industrial Alliance on SMRs and the Declaration on EU SMR 2030, and decisions on SMR deployment remain the responsibility of member states.[56] This situation may complicate efforts to articulate a cohesive SMRs policy at the EU level.

When considering the advantages and disadvantages of SMRs, it is important to prioritise efficiency, equity, safety, and democratic participation issues in their future deployment. In doing so, it prompts us to consider how to design the EU's emerging legal framework for SMRs in line with energy justice principles. As Heffron and Little[57] emphasise, delivering energy justice in the EU requires embedding justice principles into the design of regulatory and institutional processes. This approach could help to identify and mitigate the inequalities and democratic deficits that often characterise the nuclear sector.

3.3 Principles of Energy Justice in the EU's Nuclear Governance

Energy justice offers a comprehensive regulatory framework to understand and assess concerns underlying the EU's nuclear policy. This entails different forms of inequality that occur throughout the entire nuclear energy cycle, from uranium extraction to waste management and

[54] Katsiotis, N. 2025. Small Modular Reactors: Redefining global energy security for a digitalized future. *Strategy International*. Nicosia: Cyprus. Available at: https://strategyinternational.org/2025/02/05/publication161/ (Accessed: 17 May 2025).

[55] European Commission. 2023. Euratom research in action and opportunities for Europe: EU strategic autonomy and the future energy systems: EU Small Modular Reactors (SMRs) Declaration. Publications Office of the European Union. https://data.europa.eu/doi/10.2777/911262

[56] European Commission, Directorate-General for Research and Innovation. 2023. Declaration on EU SMR 2030: The role of Research, Innovation, Education and Training in the safety of Small Modular Reactors (SMRs) in the European Union. Luxembourg: Publications Office of the European Union.

[57] Heffron, R., and Little, G. F. M. 2016. *Delivering Energy Law and Policy in the EU and US: A Reader*. Edinburgh: Edinburgh University Press.

decommissioning of nuclear power plants.[58] The EU nuclear governance fails to address existing inequalities in the distribution of the benefits and costs of nuclear energy. The EU's nuclear policy prioritises large-scale, centralised projects, favouring corporate and state actors over local communities.[59] The SMRs may offer opportunities to address these inequalities due to their modularity and small size, potential reduction of GHG emissions, ability to serve weak-grid regions, potential job creation and supply chain value creation, and the involvement of Small to Medium-Sized Enterprise (SMEs) and spin-offs in the civil nuclear sector.[60]

However, SMRs are still in an early development phase and require more investment in supply chain infrastructure and regulatory adaptation. In addition, SMRs may contribute to delaying the phase-out of fossil fuels due to limited technological advancements, reinforcing inequalities in nuclear development.[61] For example, SMRs located in marginalised regions may allocate economic benefits to other areas, as their high costs require the involvement of corporate and state actors.[62] SMRs could produce up to 30 times more nuclear waste than LRs,[63] and their decommis-

[58] Heffron, R. J. and McCauley, D. 2017. The Concept of Energy Justice Across the Disciplines. *Energy Policy, 105*: 658–667. https://doi.org/10.1016/j.enpol.2017.03.018; Heffron, R. J., and McCauley, D. 2014. Achieving Sustainable Supply Chains through Energy Justice. *Applied Energy, 123*: 435–437. https://doi.org/10.1016/j.apenergy.2013.12.03

[59] To illustrate this point, one could consider the case of Italy's national radioactive waste repository. The sites identified in the CNAPI (Carta Nazionale delle Aree Potenzialmente Idonee) are primarily located in rural or marginalised regions that have not benefited from energy production but are now expected to bear its long-term environmental and social costs.

[60] European Commission, Directorate-General for Research and Innovation. 2023. Declaration on EU SMR 2030: The role of Research, Innovation, Education and Training in the safety of Small Modular Reactors (SMRs) in the European Union. Luxembourg: Publications Office of the European Union.

[61] Climate Action Network Europe. 2024. Myth Buster: Nuclear energy is a dangerous distraction. Available at: https://caneurope.org/myth-buster-nuclear-energy/ (Accessed: 23 April 2025); Goldthau, A. and Sovacool, B. K. 2012. The Uniqueness of the Energy Security, Justice, and Governance Problem. *Energy Policy, 41*: 232–240. https://doi.org/10.1016/j.enpol.2011.10.042

[62] Barnard, M. 2021. Nuclear Generation Has Huge and Hidden Security Costs. The Future Is Electric. *Medium*. Available at: https://medium.com/the-future-is-electric/nuclear-generation-has-huge-and-hidden-security-costs-e0a11264bbac (Accessed: 9 June 2025).

[63] Krall, L. M. Macfarlane, A. M. and Ewing, R. C. 2022. Nuclear Waste from Small Modular Reactors. *Proceedings of the National Academy of Sciences, 119*(23). https://doi.org/10.1073/pnas.2111833119

sioning may reveal important technical and financial uncertainties.[64] In addition, the capital costs of SMRs are 41 per cent higher than those of LRs.[65] These factors exemplify the need to strengthen safety regulations during the licensing and address the high costs of SMRs.[66]

The procedural and recognition framework emphasises the importance of transparent and participatory decision-making in the energy sector. Since the entry into force of the Aarhus Convention in 2001[67] along with its implementing legislation and the Environmental Impact Assessment Directive[68] allowing access to information, public participation, and access to justice, their application to decision-making on nuclear energy is fragmented. The role of the public also remains limited in institutional and judicial practice.[69] Civil society organisations (CISs) often engage in

[64] OECD NEA. 2024. Proceedings of the Workshop on the Management of Spent Fuel, Radioactive Waste, and Decommissioning in SMRs/ Advanced Reactor Technologies. 7–10 November 2022. Ottawa: Canada.

[65] These might include higher cost per unit of electricity compared to lLRs, potential higher operation and maintenance costs, and less competitiveness against coal and gas when carbon prices are low; Van Hee, N. *et al.* 2024. Economic Potential and Barriers of Small Modular Reactors in Europe. *Renewable and Sustainable Energy Reviews, 203.* https://doi.org/10.1016/j.rser.2024.114743

[66] Ramana, M. V., and Mian, Z. 2014. One Size Doesn't Fit All: Social priorities and technical conflicts for small modular reactors. *Energy Research & Social Science, 2*: 115–124. https://doi.org/10.1016/j.erss.2014.04.015

[67] Convention on Access to Information, Public Participation in Decision-Making and Access to Justice in Environmental Matters (Aarhus Convention), Aarhus, 25 June 1998.

[68] Directive 2003/4/EC of the European Parliament and of the Council of 28 January 2003 on public access to environmental information and repealing Council Directive 90/313/EEC, OJL 41, 14.2.2003; Regulation (EU) 2021/1767 of the European Parliament and of the Council of 6 October 2021 on the application of the provisions of the Aarhus Convention on access to information, public participation in decision-making and access to justice in environmental matters to Union institutions and bodies, OJL 356, 8.10.2021. See also: Directive 2011/92/EU of the European Parliament and of the Council of 13 December 2011 on the assessment of the effects of certain public and private projects on the environment, as amended by Directive 2014/52/EU, OJL 26, 28.1.2012; Directive 2014/52/EU of the European Parliament and of the Council of 16 April 2014 amending Directive 2011/92/EU on the assessment of the effects of certain public and private projects on the environment, OJL 124, 25.4.2014.

[69] Paks II nuclear power plant in Hungary faced criticism for limiting cross-border participation and civil society access, while public opposition to Belgium's deep geological nuclear waste repositories proposal revealed top-down level processes that inadequately considered affected communities; Puig, A. P. and Meyer, J. H. 2021. One Movement or Many? The Diversity of Antinuclear Movements in Europe. In: Kaiser, A., Lehtonen, M., Meyer, J. H. and

litigation to uphold the public's interests.[70] When such disputes require the interpretation or application of EU law, national courts or tribunals may refer questions to the Court of Justice of the EU.[71] Public participation, thus, is essential in the nuclear energy sector given its risks and technical complexity.[72][73] Thus, CECs appear to be in line with the principles of energy justice, as they could enhance participatory processes by encouraging active citizen involvement.

Cosmopolitan justice emphasises cross-border, intergenerational, and global responsibilities. As previously stated, the EU nuclear governance struggles to address shared responsibilities, such as waste management and nuclear accidents. However, some mechanisms could encourage democratic decision-making in the nuclear sector by incorporating active citizen participation and considering energy as a universal right. For example, CECs, according to the Internal Electricity Market Directive,[74] must ensure open, fair and non-discriminatory participation, particularly for low-income households and vulnerable and marginalised groups, such as elders, women, and disabled people.[75]

Rubio, M. (eds). Engaging the Atom: The History of Nuclear Energy and Society in Europe from the 1950s to the Present, Morgantown: West Virginia University Press.

[70] See, inter alia, Case C-664/15 Protect Natur-, Arten- und Landschaftsschutz Umweltorganisation v. Bezirkshauptmannschaft Gmünd,ECLI:EU:C:2017:98; Greenpeace Italy et al. v. ENI S.p.A., the Italian Ministry of Economy and Finance and Cassa Depositi e Prestiti S.p.A (2023); Case T-215/23, ClientEarth and Others v. Commission, OJ C 235, 3.7.2023.

[71] Article 267 TFEU read along with the Recommendations to national courts and tribunals in relation to the initiation of preliminary ruling proceedings, OJC C/2024/6008, 9.10.2024l; Leijon, K. 2020. It Takes Two to Tango: The preliminary reference dance between the Court of Justice of the European Union and national courts, ed. J. Krommendijk. *European Papers* 5 (2): 877–902. Available at: https://www.europeanpapers.eu/en/system/files/pdf_version/EP_eJ_2020_2_10_Articles_SS1_7_Karin_Leijon_00416.pdf. (Accessed; 11 August 2025).

[72] Tanaka Y. 2004. Major Psychological Factors Determining Public Acceptance of the Siting of Nuclear Facilities. *Journal of Applied Social Psychology, 34*: 1147–1165. https://doi.org/10.1111/j.1559-1816.2004.tb02000.x

[73] Tanaka Y. 2004. Major Psychological Factors Determining Public Acceptance of the Siting of Nuclear Facilities. *Journal of Applied Social Psychology, 34*: 1147–1165. https://doi.org/10.1111/j.1559-1816.2004.tb02000.x

[74] Article 2 (11) (a) in Directive (EU) 2019/944 of the European Parliament and of the Council of 5 June 2019 on common rules for the internal market for electricity and amending Directive 2012/27/EU, OJL 158, 14.6.2019.

[75] Koukoufikis, G. *et al.* 2023. Energy Communities and Energy Poverty. Publications Office of the European Union, Luxembourg. doi:10.2760/389514, JRC134832.

Restorative justice seeks to repair or prevent damage in instances where nuclear energy poses intergenerational risks,[76] such as, decommissioning of nuclear facilities and radioactive waste management. Integrating CECs into nuclear energy systems, for example, can help institutionalise restorative mechanisms by reinvesting in projects that deliver positive economic, environmental, and social benefits to local communities. Unlike centralised energy systems, CECs do not primarily seek to maximise financial profits. Instead, they prioritise decentralised energy facilities, active citizen participation, the redistribution of energy savings, and local job creation.[77] CECs could thus be an important mechanism for implementing energy justice in nuclear policy. However, CECs are still in early stages. Moreover, the prevailing legal framework does not permit the decentralised governance of nuclear technologies. Furthermore, nuclear power is perceived as high risk and poorly aligned with the transparency and participation principles that defined Citizens' Energy Communities (CECs). However, it is yet unknown how the member states will design future nuclear facilities in a way that prevents harm and inequalities while preserving the active participation of citizens in strategic energy decisions.[78]

Applying energy justice principles to EU nuclear governance means looking beyond technical efficiency to integrate fairness and inclusion in decision-making. This requires careful site selection for new plants or waste repositories, providing stakeholders with comprehensive information, involving them financially in projects, and assessing the compensation offered to affected communities. Local populations are often sidelined, undermining democratic legitimacy and fuelling resistance to nuclear projects. In the context of SMRs, these concerns become more pressing. Streamlined authorisation processes, meticulous safety oversight,

[76] Hazrati, M. and Heffron, R. J. 2021. Conceptualising Restorative Justice in the Energy Transition: Changing the perspectives of fossil fuels. *Energy Research & Social Science*, *78*: 102115. https://doi.org/10.1016/j.erss.2021.102115

[77] Citizen energy community entails a legal entity that '*has for its primary purpose to provide environmental, economic or social community benefits to its members or shareholders or to the local areas where it operates, rather than generating financial profits*' (emphasis added). See Article 2 (11) in Directive (EU) 2019/944; Neij, Energy communities – lessons learnt, challenges, and policy recommendations, *Oxford Open Energy*, 2025, 4.

[78] Reins, L., and Wijns, J. 2025. The 'Safe and Sustainable by Design' Concept - A regulatory approach for a more sustainable circular economy in the European Union? *European Journal of Risk Regulation*,16: 96–113. https://doi.org/10.1017/err.2024.29; IAEA. 2023. Decommissioning of Nuclear Installations. Vienna: IAEA. Available at: https://www.iaea.org/topics/decommissioning (Accessed: 11 August 2025).

and economic feasibility evaluations are imperative to ensure their deployment aligns with energy justice.[79]

3.4 Concluding Remarks

The Russian–Ukrainian War has exposed the EU's energy vulnerability, prompting several EU member states to reconsider nuclear power as a strategic option for achieving energy security and decarbonisation. Given its low carbon emissions and its capacity to complement renewable energy sources, nuclear energy represents a strategic resource to ensure the EU's continuous and reliable energy supply. This chapter demonstrates that the existing EU's nuclear governance system fails to tackle several challenges associated with the development of nuclear power, ensuring transparent decision-making, access to information and justice in the event of grievances associated with nuclear development. In addition, it is necessary to identify, analyse, and assess multiple inequalities and environmental impacts throughout the nuclear power development life cycle. SMRs and CECs could offer opportunities to strengthen public participation in decision-making and promote citizens' engagement in community-based energy initiatives. However, questions persist about the long-term economic viability, waste management, social awareness and legal framework for these systems. The absence of coordinated nuclear policy among EU member states remains a significant gap. Our research findings suggest that the EU should coordinate a better political and legal dialogue through multi-level governance to critically assess the role of nuclear energy in its energy transition in line with the principles of energy justice. Thus, integrating an energy justice perspective into the EU's nuclear policy seeks to rethink and redefine the governance of nuclear energy. It invites reflections on how the EU navigates the complex interdependencies and contradictions between its energy security and climate objectives while ensuring that social impacts and justice considerations remain cornerstones in its legal and policy framework.

[79] Heffron, R. J. 2024. Energy Justice - The First Step in an Energy Decision Today. In: Heffron, R. J. and de Fontenelle, L. (eds). *The Power of Energy Justice & the Social Contract*. Cham: Palgrave Macmillian, 7. https://doi.org/10.1007/978-3-031-46282-5_1; Tuler, S., and Webler, T. 2024. The Challenge of Community Acceptance of Small Modular Reactors. *Energy Research & Social Science*, *118*: 1. https://doi.org/10.1016/j.erss.2024.103831

CHAPTER 4

Regenerating Cities Through Energy Communities: A Justice-based Comparative Perspective from Roma (Italy) and Brasilia (Brazil)

Ari Rogério Ferra Júnior and Gianna Giardini

Abstract This chapter uses two case studies – the Le Vele Renewable Energy Community in Roma and the Solidarity Energy (*Energia Solidária*) Project in Brasília – to explore how renewable energy communities (RECs) can act as catalysts for urban regeneration and energy justice. Taking an interdisciplinary approach, the analysis combines legal and policy frameworks with empirical evidence to examine the intersection between energy transition, social inclusion and urban transformation. In Roma, the Le Vele initiative shows how a community-based photovoltaic system can cut energy costs, reduce carbon emissions and encourage social equity in a disadvantaged urban area. Backed by a foundation committed to tackling energy poverty, the project showcases the potential of cooperative energy models to foster distributive,

A. Rogério Ferra Júnior
PhD in Civil Law from the University of São Paulo
and the University of Camerino, Camerino, Italy
e-mail: ariferrajr@gmail.com

G. Giardini
Istituto Universitario di Studi Superiori Università di Pavia
and Università di Camerino, Camerino, Italy

A. Phillips, L. Kaschny (eds.), *Global Perspectives on the Circular Economy in the Just Transition*, Just Transitions,
https://doi.org/10.1007/978-3-032-18342-2_4

procedural and recognition justice. Revenues from shared energy are reinvested in welfare services, thereby reinforcing solidarity and community resilience. In Brazil, the Solidarity Energy Project reflects a similar pursuit of an inclusive energy transition within a distinct regulatory and socio-economic context. Despite the absence of a legal framework for RECs in Brazil, the initiative combines photovoltaic installation, professional training and social tariff mechanisms to increase access to sustainable energy for low-income families. By integrating the five dimensions of energy justice – distributive, procedural, recognition, restorative and cosmopolitan – the projects help to reduce inequalities and advance Sustainable Development Goals 7, 10, 11 and 13. A comparative analysis reveals structural similarities and contextual differences between the two cases. While Italy benefits from a stable regulatory framework aligned with EU directives, Brazil excels in linking energy transition to social inclusion and employment creation. Together, these experiences demonstrate that RECs and solidarity-based energy projects can regenerate urban environments, empower citizens and promote equitable access to energy. Integrating energy justice into urban regeneration provides a transformative approach to creating sustainable, inclusive and resilient cities.

Keywords Cities • Urban and Social Regeneration • Energy Poverty • Renewable Energy Communities • Ecology of Proximity • SDG 7 • Energy Justice

4.1 Introduction

Today, nearly 70 per cent of the global population lives in urban areas.[1] This figure is expected to increase in the future, given that cities are the main centres of political, economic and social life. Urban regeneration is born as

[1] In the European Union, 39% of the EU population lives in large urban centres, 36% in medium-sized towns and suburbs and 25% in rural areas. European Union. (2023). *Facts and figures on the European Union*. European Union. Available at: https://european-union.europa.eu/principles-countries-history/facts-and-figures-european-union_en (Accessed: 20 August 2025). In Brazil, 87% of the Brazilian population lives in urban areas. In Brazil, the majority of the population (approximately 87.4%) lives in urban areas, while a smaller percentage (12.6%) resides in rural areas, according to the 2022 Population Census. Specifically, 36% live in medium-sized towns and suburbs, and 25% live in rural areas. Brazil's urbanisation rate is quite high, with a significant portion of the population concentrated in large urban centres, particularly in regions like São Paulo, Rio de Janeiro and Brasília. IBGE – News Agency. (2023). 2022 Census: 87% of the Brazilian population lives in urban areas. IBGE News Agency. Available at: https://agenciadenoticias.ibge.gov.br/en/agencia-news

a response to uncontrolled urban growth and the subsequent degradation that results from it.[2] The aim of urban regeneration[3] is to address problems of physical decay and socio-economic distress in an integrated way,[4] offering a comprehensive opportunity to rethink, redesign and modernise cities to make them fit for a green, digital and inclusive society, while supporting economic growth.[5] From this perspective, urban regeneration is one of the strategic levers for achieving the United Nations Sustainable Development Goals (SDGs).[6] Initially focused solely on the physical transformation of spaces, the approach has evolved to become multidimensional and holistic, integrating social, environmental and economic aspects. These processes require cross-cutting knowledge and skills in areas such as urban planning, landscape and environment, sustainability and the law.[7] Furthermore,

/2184-news-agency/news/41910-2022-census-87-of-the-brazilian-population-lives-in-urban-areas (Accessed: 30 July 2025).

[2] Said, Z.M. and Dindar, S. 2024. Key Challenges and Strategies in the Evaluation of Sustainable Urban Regeneration Projects: Insights from a Systematic Literature Review. *Sustainability*: *16*(22). https://doi.org/10.3390/su16229903

[3] Martins, M. L. R., & Santos Pereira, A. L. D. 2019. Urban Regeneration in the Brazilian urban policy agenda. *European Planning Studies*, *27*(6), 1129–1145. https://doi.org/10.1080/09654313.2019.1598021

[4] Urban regeneration is a widely recognised concept, but without a single definition, which has evolved from a conservative vision to an innovative and integrated approach, being influenced by judges and legislators. Regional Administrative Court, Puglia Region. (2019). *Judgment of 13 December 2019, No. 1967.* In Brazil, the Federal Supreme Court (STF), although it has not ruled specifically on urban regeneration, has strengthened the issue by validating the Statute of the Metropolis (Law No. 13,089/2015) in the judgment of Direct Action of Unconstitutionality (ADI) 5857, which requires integrated and sustainable urban development plans, in addition to recognising the autonomy of municipalities to define land use, including in green areas, facilitating social housing and urban revitalisation policies. ADI 5857 is a Direct Action of Unconstitutionality filed with the Federal Supreme Court (STF) by the government of Pará, challenging provisions of the Statute of the Metropolis (Law 13.089/2015). The main focus of the action is the constitutionality of the requirement for an Integrated Urban Development Plan (PDUI) prepared by representatives of the state, municipalities and civil society, based on the principle of cooperative federalism.

[5] Scuto, L., 2022. The development of a new model of urban regeneration in light of the digitalization of public administration. *CERIDAP*, 2. https://doi.org/10.13130/2723-9195/2022-3-17

[6] UN-Habitat Flagship Programme. 2024. Urban regeneration for localizing the Sustainable Development Goals. UN-Habitat. Available at: https://unhabitat.org/sites/default/files/2024/05/ur_-_paper_22052024.pdf (Accessed: 19 Aug 2025).

[7] Said, Z.M. and Dindar, S. 2024. Key Challenges and Strategies in the Evaluation of Sustainable Urban Regeneration Projects: Insights from a Systematic Literature Review. *Sustainability*: *16*(22). https://doi.org/10.3390/su16229903

designing and transforming urban spaces requires collaboration between the public, private and third sectors. In the *One Health* approach era, the public health sector must not be left out of these developments.[8] Scientific evidence demonstrates a correlation between environmental characteristics and physical and mental health outcomes.[9] Nevertheless, urban regeneration is integral to climate change adaptation and mitigation. Buildings and infrastructure must be designed to withstand extreme weather events and use low-carbon materials.[10] Finally, from a social perspective, sustainable regeneration should strengthen urban equity, rather than leading to gentrification or exclusion.[11]

In the era of energy transition, urban regeneration cannot ignore the need to strengthen the relationship with energy and environmental policies.[12] When we talk about making cities more sustainable, energy services are a priority. Many scholars have emphasised the importance of reducing energy consumption and emissions that contribute to climate change by renovating existing buildings.[13] These interventions aim to enhance the energy efficiency of buildings by encouraging the adoption of advanced technologies and renewable energy sources to minimise energy demand. The adoption of smart grids and digital systems for managing consumption also enables the distribution and use of energy to be optimised more efficiently, making cities more resilient and autonomous.[14] Experiences of

[8] Le Gouais, A. and Cooper, M. 2024. Who can influence healthy urban development? *Cities & Health, 9*(4): 644–656. https://doi.org/10.1080/23748834.2024.2362508

[9] World Health Organization. Regional Office for Europe. 2016. World Health Organization. Regional Office for Europe. Available at: https://iris.who.int (Accessed: 20 August 2025).

[10] Said, Z.M. and Dindar, S. 2024. Key Challenges and Strategies in the Evaluation of Sustainable Urban Regeneration Projects: Insights from a Systematic Literature Review. *Sustainability*: *16*(22). https://doi.org/10.3390/su16229903

[11] Dindar, S. 2025. A Systematic Review of Urban Regeneration's Impact on Sustainable Transport: Traffic Dynamics, Policy Responses, and Environmental Implications. *Sustainable Development*. https://doi.org/10.1002/sd.70007

[12] Balletto, G. *et al.* 2022. More Circular City in the Energy and Ecological Transition: A Methodological Approach to Sustainable Urban Regeneration. *Sustainability*:14 (22). https://doi.org/10.3390/su142214995

[13] Razzaq I., *et al.* 2022. Reduction in energy consumption and CO2 emissions by retrofitting an existing building to a net zero energy building for the implementation of SDGs 7 and 13. *Frontiers in Environmental Science, 10*. https://doi.org/10.3389/fenvs.2022.1028793

[14] Khaleel, M., *et al.* 2024. Impact of Smart Grid Technologies on Sustainable Urban Development. *International Journal of Electrical Engineering and Sustainability (IJEES)*, *2*(2), 62–82. https://doi.org/10.5281/zenodo.11577746

energy *prosumagerism,*[15] such as self-consumption and energy communities, are becoming an integral part of regeneration processes, offering citizens the opportunity to participate in activities of general interest, such as tackling energy poverty.[16] According to the most recent data, 9.2% of people in the European Union (EU) are affected by energy poverty (Eurostat, 2024),[17] meaning almost one in ten EU citizens cannot meet their energy needs. This has serious repercussions for people's health, well-being and quality of life. In our case study of Italy, the Italian Observatory on Energy Poverty (OIPE) found that 2.36 million households were living in energy poverty in 2023, accounting for 9% of the total. This is the highest level since the beginning of the statistical series and represents an increase of 1.3 percentage points compared to 2022, equating to approximately 340,000 more households.[18] In Brazil, energy poverty still affects 11% of Brazilian households.[19] Energy poverty remains a paradox between low income and increased household energy consumption in Brazil, as the government's subsidy is (apparently) unattainable for the low-income

[15] Prosumagerism represents a paradigm shift in which energy consumers, empowered by advancements in digital technologies and renewable energy sources, take on the roles of producers of energy resources. This model marks a transformation from centralised to decentralised systems, in which individuals can generate, share, store and trade energy. This gives them a central role in governing the electricity market and contributes to the democratisation of the system. On this point see, among the others, Ruggeri, L. 2022. Consumatore e prosumerismo energetico nel quadro regolatorio europeo. Actualidad Jurídica Iberoamericana, 16(bis), 415–436; European Commission. 2017. Residential prosumers in the European Energy Union. Publications Office of the European Union. Available at: https://commission.europa.eu/system/files/2017-11/study-residential-prosumers-energy-union_en.pdf (Accessed: 2 September 2025).

[16] Energy poverty is defined as the condition in which households lack the necessary resources to ensure an adequate level of heating, cooling, lighting and other essential energy services. In argument, see Joint Research Centre (JRC), European Commission. 2023. Energy Communities and Energy Poverty — The Role of Energy Communities in Alleviating Energy Poverty. JRC Technical Report.

[17] European Commission. Directorate-General for Energy. 2024. Energy poverty. European Commission. Available at: https://energy.ec.europa.eu/topics/markets-and-consumers/energy-consumers-and-prosumers/energy-poverty_en?utm_source=chatgpt.com (Accessed: 19 August 2025).

[18] OIPE – Osservatorio Italiano sulla Povertà Energetica. 2024. La povertà energetica in Italia nel 2023. Centro Levi-Cases, University of Padova.

[19] Bezerra, P. *et al.* 2022. The multidimensionality of energy poverty in Brazil: A historical analysis. *Energy Policy, 171.* https://doi.org/10.1016/j.enpol.2022.113268

population. So much so that the government introduced new social energy tariffs in 2025.[20]

In light of the considerations, the following chapter examines how urban regeneration can promote energy justice and ensure fair and inclusive access to energy in cities. It adopts an interdisciplinary approach that integrates legal and political analysis with empirical research to investigate the relationship between urban regeneration and energy justice. The legal analysis focuses on the main regulatory instruments and the surrounding doctrinal interpretations, while the empirical component draws on two case studies located in areas particularly affected by energy poverty and unequal access to energy. The first case examines an urban regeneration initiative in Roma, and the second focuses on the city of Brasília, Brazil. The research is based on a review of institutional documents, technical reports, urban planning instruments and policy papers relevant to each context. Where possible, this documentary analysis is complemented by semi-structured interviews with professionals from the sectors directly involved in the urban regeneration processes.

Section 2 examines the Le Vele Renewable Energy Community (REC) in Roma, considering it in the context of the SDGs' framework for sustainable and inclusive urban regeneration. In this context, energy communities are viewed as key tools for tackling energy poverty and strengthening civic participation and social cohesion.

Section 3 shows that, despite the lack of a specific regulatory framework in Brazil to promote renewable energy communities (RECs) in comparison to the European Union, the country has made progress in its commitments to the 2030 Agenda, especially United Nations Sustainable Development Goal 7 (SDG 7), seeking to ensure universal and affordable access to energy. Despite widespread electricity coverage, challenges persist, such as high tariffs, regional inequalities and barriers faced by vulnerable populations in accessing clean energy sources. In this context, the Solidarity Energy (*Energia Solidária*) Project, developed in the Federal District, stands out as an initiative that combines social inclusion and energy transition through the installation of 1,000 photovoltaic kits in low-income homes and the technical training of 1,300 or more people

[20] Simões, GM F. and Leder, S. M. 202. Energy poverty: The paradox between low income and increasing household energy consumption in Brazil. *Energy and Buildings*, *268*. https://doi.org/10.1016/j.enbuild.2022.112234

in solar energy.[21] The project promotes access to the Social Electricity Tariff,[22] fosters income generation and contributes to environmental justice and the reduction of inequalities, integrating sustainability, education and community participation.

Section 4 provides a comparative analysis between the Le Vele REC and the Solidarity Energy (*Energia Solidária*) community energy project. The goal is to identify the differences, challenges and gaps in both models, providing a broader understanding of the practices and impacts of sustainable energy initiatives in different community contexts. The aim of the chapter is to demonstrate how renewable energy and RECs can stimulate socio-urban regeneration through solidarity and innovation.

4.2 Energy Communities as Key Enablers of Urban Regeneration

4.2.1 The MLa Vele REC Case (Italy)

The SDGs emphasise the importance of creating more sustainable cities and communities through the promotion of responsible consumption and production practices (SDGs 11 and 12).[23] Cities should ensure that everyone has access to essential services, energy, housing, transport and green

[21] Ministry of Science, Technology and Innovation; Avante Brasil (OSCIP). 2024. Work Plan – Energia Solidária Project [Project funded through a fostering agreement – Process No. 01245.000351/2024-96]. Available at: https://avantebrasil.ong.br/wp-content/uploads/2025/01/Plano_de_Trabalho_-_PRENERGIS_assinado.pdf (Accessed:14 August 2025).

[22] The Social Electricity Tariff is a public policy offering discounts on electricity bills for: low-income families and for the elderly or people with disabilities who receive the Continuous Social Assistance Benefit. This public policy is defined by Law No. 12,212 of 20 January 2010, Law No. 10,438 of 26 April 2002 and Interministerial Ordinance MME/MS No. 630 of November 8, 2011. The National Electric Energy Agency regulates the application of the discounts in Articles 176 to 179 and Article 200 of Normative Resolution No. 1,000/2021. Recently, Provisional Measure No. 1,300/2025, dated 21 May 2025, amended Law No. 12,212/2010, establishing a new rule for discounts on bills issued from July 5, 2025, onwards. Under the new rules, families eligible for the benefit will not have to pay for the first 80 kilowatt-hours (kWh) consumed each month. *Social Electricity Tariff.* 2025. Available at: https://www.gov.br/aneel/en/subjects/tariffs/social-tariff (Accessed: 31 July 2025).

[23] United Nations. 2015. Transforming our world: The 2030 Agenda for Sustainable Development. New York: United Nations.

public spaces, while improving resource use and reducing environmental impact.[24] The focus on social justice and reducing inequalities (Goal 10) stresses the link between urban renewal and social inclusion.[25] To this end, the EU actively supports civic participation in urban regeneration processes to strengthen collective awareness and shared action.[26]

The experience of the energy community Le Vele in Roma (IT) fits into this scenario. Established in 2022, the REC[27] Le Vele[28] was born in

[24] Elmqvist, T., *et al.* (eds.). 2015. *Urbanization, biodiversity and ecosystem services: Challenges and opportunities*. Springer. https://doi.org/10.1007/978-94-007-7088-1

[25] Michaels, R., *et al.* 2021. *The private side of transforming our world: UN Sustainable Development Goals 2030 and the role of private international law*. (1st ed.). Intersentia studies on Private International Law.

[26] Healey, P. 2010. *Making better places: The planning project in the twenty-first century*. Palgrave Macmillan; European Commission. 2016. *Urban Agenda for the EU: Pact of Amsterdam*. Brussels: European Union.

[27] Renewable Energy Communities (RECs) are regulated by Article 22 of Directive (EU) 2018/2001 (RED II), which defines their legal status, objectives and operational scope. An energy community is an autonomous legal entity comprising citizens, local authorities and small enterprises that collaborate to produce, consume, share, store and sell energy from renewable sources. According to the European legislator, RECs are guided by the principles of open and voluntary participation combined with effective control by members and shareholders, provided that the participating enterprises' main commercial activity is not energy-related. The RED II Directive emphasises that renewable energy communities are not primarily profit-driven but rather aim to generate environmental, economic and social benefits for their members and the local community. Furthermore, EU law requires RECs to be independent of individual members and traditional energy market operators to ensure strategic decisions reflect community interests rather than those of dominant market actors. Article 22 of the RED II Directive allows renewable energy communities to operate across the entire energy value chain, from production and distribution to supply, storage, sharing and providing related services. Thus, RECs play a pivotal role in the European energy transition by promoting inclusive and participatory governance that combines environmental objectives with the principles of social and economic pluralism. The European legislative framework distinguishes between Citizen Energy Communities (CECs) and Renewable Energy Communities (RECs). While both types of community aim to promote local participation in the energy transition, CECs (Directive (EU) 2019/944) have a broader scope. They can operate across the electricity sector and are not limited to renewable sources. Participation is open to all citizens, local authorities and small enterprises, regardless of geographical proximity. In contrast, RECs (Directive (EU) 2018/2001) are specifically linked to renewable energy generation, requiring proximity between members and projects to ensure that energy production and its benefits are tied to the surrounding community. In the case of Le Vele, the REC model was chosen as the most appropriate option since the renewable source involved is solar energy and the principle of geographical proximity applies.

[28] CER Le Vele. 2025. Available at: https://www.cerlevele.it (Accessed: 31 July 2025).

response to high energy demand in an area of significant social vulnerability. Promoted by the non-profit organisation Leonarda Vaccari Institute, together with partners such as the Istituto Comprensivo Claudio Abbado, the association VAILV Amici dell'Istituto, Il Vento Sulla Vela and Scenario B Aps, the project involved the shared production and consumption of electricity from renewable sources, with the aim of saving energy, reducing emissions and implementing social equity.

The Leonarda Vaccari Institute[29] is a centre of excellence for rehabilitation and social education, welcoming over 5,000 people with physical and cognitive disabilities each year. The institute offers personalised services in a highly specialised environment. The facility hosts a primary school with seven classes where teaching is supported by technological aids, multi-sensory activities and trained staff. It also provides outpatient rehabilitation for adults and children, including speech therapy, physiotherapy and psychomotricity. For people with more complex needs, the institute offers semi-residential services and a residential unit for chronic or post-acute patients. It also includes a regional centre for the evaluation and customisation of technological and computer-based aids to support communication and independence. The facilities feature a multi-sensory room for children with severe disabilities, medicalised care areas, therapy rooms, gyms, laboratories, technical offices and shared spaces. All of these functions make the Institute highly energy intensive. Consumption comes from electronic devices, specialised lighting and the continuous use of medical equipment.

The rise in utility costs – driven by the pandemic and the war in Ukraine – has significantly affected the financial stability of the Vaccari Institute. In response, the Municipality of Roma stepped in as a facilitator to help establish a renewable energy community. The financing and permitting process was particularly complex.[30] From a financial perspective, the initiative was made possible due to the support of a foundation dedicated to addressing energy poverty, which covered the entire cost of the photovoltaic system. In addition, a specialised technical office provided *pro bono* expertise on engineering and legal matters. From an administrative

[29] Istituto Leonarda Vaccari. 2025. Available at: https://www.leonardavaccari.it (Accessed:31 July 2025).

[30] This analysis draws on insights from two semi-structured interviews with professionals directly involved in the project. Both interviewees were responsible for project management and implementation activities.

standpoint, the authorisation process was highly bureaucratic, given that the intervention was located in Roma's historic centre and involved a listed building. Nevertheless, the system's physical configuration – installed on a flat, non-visible surface and fully integrated into the architectural context – facilitated the project's eventual approval. The energy community has been operational since 2024.[31] The installed system has a total power output of 82 kW and consists of 198 modules. It is capable of producing around 120,000 kilowatt-hours per year. It saves around 41 tonnes of CO_2 per year, which is equivalent to planting 1,365 trees. This makes a concrete contribution to the ecological transition and the fight against energy poverty.[32] In 2024, the system produced around 98 megawatt-hours, of which approximately 66 megawatt-hours were self-consumed by the group (67.8%), with the remainder being exported (32.2%).[33] The REC's full digitalisation makes it possible to collect data, enabling actual energy consumption to be measured.[34]The shared energy and incentives obtained are used entirely for social purposes. According to an internal regulation, 100% of the economic contribution generated by energy sharing is donated to those experiencing economic or social difficulties or used to provide additional services that the Vaccari Institute could not guarantee alone. Thus, the energy produced is returned to the community, supporting vulnerable individuals and investing in welfare. Training aimed at enhancing energy awareness and providing the necessary skills for autonomous energy community management has also been promoted.[35]

In a context where energy is synonymous with life – particularly for those who depend on machines to communicate or survive – Le Vele is a tangible example of a community caring for the most vulnerable through innovation and solidarity. Its photovoltaic system reduces costs and restores

[31] These insights are based on primary qualitative data collected through semi-structured interviews with professionals directly involved in the *Le Vele REC* project.

[32] Barja-Martínez, S., *et al.* 2025. Local energy communities optimization considering cost and greenhouse gases minimization. https://arxiv.org/abs/2504.20973

[33] CER Le Vele. Analisi dei Dati Energetici 2024. Available at: https://www.cerlevele.it (Accessed: 31 July 2025).

[34] Bonavolontà, F., *et al.* 2025. Real-Time Monitoring of Energy Contributions in Renewable Energy Communities Through an IoT Measurement System. *Sensors*, *25*(5). https://doi.org/10.3390/s25051402

[35] Rinnovabili. 2025. Available at: https://www.rinnovabili.it/energia/comunita-energetiche-rinnovabili/cer-le-vele-lenergia-rinnovabile-illumina-il-futuro-sociale-di-roma/ (Accessed: 31 July 2025).

dignity and hope to people with disabilities and those living in marginalised communities. Inspired by the Le Vele model, a municipal Regulation[36] has been created to govern the use of public areas and facilities for the creation of energy communities within the third sector through shared administration.[37] Roma collaborates with active citizens and third-sector entities (Ente del Terzo Settore – ETS) to manage and enhance shared resources (such as installations and photovoltaic areas) based on a principle of trust.[38] This shared administration involves a co-planning stage, during which the administration and the ETSs jointly define the energy and social needs (e.g. combatting energy poverty) and identify shared goals to inform interventions. The Administration and the CERs (comprising ETSs) then develop projects together to produce and share energy from renewable sources, build plants and redistribute the benefits to the local community via conventions and collaboration agreements.

The Le Vele REC experience offers some key lessons. Firstly, although the topic has not yet been widely explored in international law, the connection between energy and human rights is becoming clearer. Access to energy is essential for the full realisation of human rights. It is a vital component in meeting basic human needs, such as access to clean water, healthcare, food, housing and transport.[39] Despite its importance, major international human rights treaties do not explicitly recognise access to energy as a fundamental right in its own right. Nevertheless, many scholars[40] argue that access to energy is a prerequisite for the effective enjoyment of recognised fundamental rights such as the right to life, health and an adequate standard of living. From this perspective, energy is considered a derivative right, that is a right that is necessary for the full realisation of other human rights. Even the United Nations emphasises the

[36] Comune di Roma. 2025. Regolamento CER Assemblea Capitolina. Available at: https://www.comune.roma.it/webresources/cms/documents/Mun11_Regolamento_CER_AssembleaCapitolina_1.pdf (Accessed: 20 August 2025).

[37] Italy. 2017. Legislative Decree No. 117 of 3 July 2017, Code of the Third Sector, Article 55.

[38] Italy. 2023. Legislative Decree No. 36 of 31 March 2023, Public Contracts Code, Article 2 (Principle of Trust).

[39] Wewerinke-Singh, M. 2021. A human rights approach to energy: Realizing the rights of billions within ecological limits. *Review of European, Comparative & International Environmental Law*, *31*(1), 16–26. https://doi.org/10.1111/reel.12412

[40] Löfquist, L. 2020. Is there a universal human right to electricity? *The International Journal of Human Rights*, *24*(6): 711–723. https://doi.org/10.1080/13642987.2019.1671355

importance of energy in ensuring a dignified life through the UN SDG 7. Universal access to energy is closely linked to fighting poverty and reducing inequalities: without it, millions of people are excluded from adequate living conditions.[41]

Secondly, energy communities have proven to be powerful catalysts for socio-ecological regeneration.[42] They introduce a new planning model that puts citizens at the heart of the regeneration process and encourages them to play an active role in urban policy. Instead of remaining passive consumers, citizens become active participants in the generation, distribution and management of energy.[43] This shift from inactivity to energy activism not only strengthens local energy autonomy, but also fosters greater social and environmental awareness.[44] Furthermore, citizens decide how to allocate revenues, choosing whether to reinvest them in improving local infrastructure or providing economic support for the most vulnerable members of the community, thereby promoting equitable access to energy and reducing inequalities.[45] The result is an integrated vision of urban regeneration combining environmental sustainability, social inclusion and economic and technological innovation, achieved through collaborative involvement between public administrations and private organisations.

Methodologically, our analysis is grounded in the energy justice framework.[46] This allows us to view energy communities as instruments of

[41] Grossmann, K., and Trubina, E. 2021. How the Concept of Dignity Is Relevant to the Study of Energy Poverty and Energy Justice. *Frontiers in Sustainable Cities, 3*, 1–11. https://doi.org/10.3389/frsc.2021.644231

[42] Cocco, A. 2022. The Role of Energy Communities in the Energy Transition. *The Italian Law Journal, 2*, 509-525. https://doi.org/10.23815/2421-2156.ITALJ; D'Alpaos, C., & Andreolli, F. 2021. Renewable energy communities: The challenge for new policy and regulatory framework design. *New metropolitan perspectives: knowledge dynamics and innovation-driven policies towards urban and regional transition, 2*, 500–509. Di Lorenzo, G., *et al.* 2021. Innovative power-sharing model for buildings and energy communities. *Renewable Energy, 172*: 1087–1102. https://doi.org/10.1016/j.renene.2021.03.063

[43] Neves, C.; Oliveira, T. and Sarker, S. 2025. Citizens' participation in local energy communities: The role of technology as a stimulus. *European Journal of Information Systems, 34*(1): 122–145. https://doi.org/10.1080/0960085X.2024.2302426

[44] Neij, L. *et al.* 2025. Energy communities—lessons learnt, challenges, and policy recommendations. *Oxford Open Energy, 4*. https://doi.org/10.1093/ooenergy/oiaf002

[45] Campagna, L., *et al.* 2024. Renewable energy communities and mitigation of energy poverty: Instruments for policymakers and community managers. *Sustainable Energy, Grids and Networks, 39*. https://doi.org/10.1016/j.segan.2024.101471

[46] McCauley, D., *et al.* 2013. Advancing energy justice: The triumvirate of tenets. *International Energy Law Review, 32*(3), 107–116; Heffron, R. J., & McCauley, D. 2017. The

socio-urban regeneration, connecting distributive, procedural, recognition and cosmopolitan justice with the wider objectives of equity, sustainability and inclusion, rather than considering them solely as technical infrastructures. In this regard, the Le Vele REC is a tangible manifestation of energy justice. Firstly, it ensures the equitable distribution of energy resources among citizens, reducing disparities in access and promoting distributive justice. Secondly, the cooperative model that underpins the REC promotes transparency and citizen participation in decision-making processes, thereby reinforcing procedural justice and fostering trust between institutions and citizens. Thirdly, by actively including marginalised and vulnerable groups within its governance structure, the REC promotes recognition justice and counters the exclusion that is often characteristic of conventional energy systems.[47] Finally, by promoting affordable, universal access to renewable energy, the initiative strengthens cosmopolitan justice, linking local benefits to broader sustainability and equity objectives. Taken together, these dimensions demonstrate that the Le Vele REC is a socially transformative instrument capable of enhancing community resilience and driving urban regeneration.

More precisely, Le Vele REC is a clear example of energy justice in action.[48] The directive (EU) 2018/2001 provides the necessary legislative framework to support the development of RECs, guaranteeing legal certainty for their members.[49] However, financial and administrative barriers still hinder their wider adoption.[50] Le Vele REC's success demonstrates how these obstacles can be overcome through social innovation and strong local partnerships. It shows that legislation and grassroots engagement can transform structural challenges into opportunities for inclusive and sustainable cities and urban regeneration.

concept of energy justice across the disciplines. *Energy Policy, 105*: 658–667. https://doi.org/10.1016/j.enpol.2017.03.018

[47] Heffron, R. J. 2023. Energy justice – the triumvirate of tenets revisited and revised. *Journal of Energy & Natural Resources Law*, *42*(2): 227–233. https://doi.org/10.1080/02646811.2023.2256593

[48] Sovacool, B. K., *et al.* 2016. Energy decisions reframed as justice and ethical concerns. *Nature Energy*, *1*. https://doi.org/10.1038/nenergy.2016.24

[49] Ruggeri, L. 2021. Legislative policies and jurisprudence on climate change: New tools for removing the barriers to new forms of energy consumption. In: Ruggeri, L. (Ed.), *Needs and Barriers of Prosumerism in the Energy Transition Era* (pp. 98–109). Dykinson.

[50] Caramizaru, A., and Uihlein, A. 2020. *Energy communities: An overview of energy and social innovation* (EUR 30083 EN). Luxembourg: Publications Office of the European Union.

4.2.2 *The Solidarity Energy Project (Brazil)*

Despite the differences between the two regulatory systems, especially regarding laws addressing energy communities, the context of energy poverty (definition, data, indicators), the regulatory and institutional framework (laws, policies, programs) and specific projects/experiences, Brazil has different examples of energy communities or 'energy in the communities'. Firstly, Brazil does not have legal instruments specifically aimed at promoting communities and the use of renewable energy in the same way as the European Union.[51] But there is no renewable energy community in Brazil following the EU model, but there is energy present in communities with a social focus.[52] Law 14,300/2022 establishes the Electric Energy Compensation System, which is a net metering mechanism.[53]

The promotion of clean and sustainable energy sources is essential for realising the principle of sustainable development and for reducing dependence on conventional sources, mainly based on fossil fuels. However, many social groups, particularly those in situations of socio-economic vulnerability, face structural barriers in accessing such technologies. This highlights the need for inclusive public policies and regulatory mechanisms that ensure the right to equitable access to sustainable energy, in accordance with the principles of environmental justice and human dignity. By highlighting the benefits of rural electrification, the installation of photovoltaic systems in low-income housing and energy efficiency in social projects, these studies offer relevant contributions to the formulation of public policies at the state and federal levels. It is recognised that investment in renewable energy goes beyond poverty alleviation, also boosting sustainable development and the resilience of the most vulnerable communities. In this context, policies that encourage and enable the adoption of these technologies have the potential to play a strategic role

[51] Iglesias, C., and Vilaça, P. 2022. On the regulation of solar distributed generation in Brazil: A look at both sides. *Energy Policy,* 167. https://doi.org/10.1016/j.enpol.2022.113091

[52] Hochstetler, R. L., and Born, P. H. S. 2022. Community energy design models in Brazil: From niches to mainstream. In: S. Löbbe, F. Sioshansi, & D. Robinson (Eds.), *Energy Communities: Customer-Centered, Market-Driven, Welfare-Enhancing?* (pp. 317–338). Academic Press.

[53] de Souza, A. *et al.* 2023. A comparison study of grid impact of photovoltaic installations in Brazil according to Normative Resolution 482 and Federal law 14.300. *Social Science Research Network.* https://dx.doi.org/10.2139/ssrn.4309470

in overcoming the country's socio-economic challenges.[54] This aligns with Brazil's fundamental objectives, which include building a free, just and united society; guaranteeing national development; eradicating poverty and marginalisation; reducing social and regional inequalities; and promoting the well-being of all, without prejudice of origin, race, gender, colour, age, or any other forms of discrimination, as set out in Article 3 of the Constitution of the Federative Republic of Brazil.[55]

What stands out in the current Brazilian energy transition context is that, despite Brazil's significant progress in universalising access to electricity and in the substantial share of renewable sources in its energy matrix, electricity tariffs remain among the highest in the world. This compromises effective accessibility, especially for economically vulnerable segments of the population. The combination of high taxes, transmission costs and sectoral charges contributes to the rising cost of this bill. Brazil has a tariff flag system[56] that influences electricity consumption in Brazil, as tariff flags represent a demand-side management[57] mechanism used in the Brazilian electricity sector and act as a signal sent to agents about the costs of electricity production. Power generation in Brazil relies heavily on rainfall, as the electricity matrix is predominantly hydroelectric.[58]

[54] da Silva, E. C. M. *et al.* 2024. The Role of Renewable Energies in Combating Poverty in Brazil: A Systematic Review. *Sustainability*, *16*(13). https://doi.org/10.3390/su16135584

[55] De Cicco, M. C., and Ferra Júnior, A. R. 2022. Fundamental duty to sustainable development in the Brazilian Federal Constitution of 1988 and solar energy (SDG 7) as an energy policy in the current Brazilian conjuncture. In: L. Ruggeri & F. Pascucci (Eds.), *Prosumerism and energy communities: Expanding concepts in a global perspective* (pp. 75–83). Scholarly Society Publisher of Wien.

[56] Tariff flags are the system that signals to consumers the real costs of electricity generation. To this end, the flag colours (green, yellow, or red) indicate whether the energy will cost more or less depending on the electricity generation conditions. *About Tariff Flags.* 2025. Available at: https://www.gov.br/aneel/pt-br/assuntos/tarifas/bandeiras-tarifarias (Accessed: 14 August 2025).

[57] In Brazil, demand management is primarily carried out through the Tariff Flag System, a mechanism created by the National Electric Energy Agency. This system uses colours on monthly energy bills to signal to consumers the real costs of electricity generation. The idea is that, by indicating variations in these costs—often impacted by factors such as lack of rainfall, which compromises hydroelectric power generation—the system encourages more conscious consumption. Although its main function is to anticipate revenues for distributors and reflect the immediate costs of generation, the Tariff Flag System also acts as a form of demand response, providing information that helps consumers adjust their consumption habits.

[58] The tariff flags are green: normal production conditions; yellow: less favorable situation, generating an increase in the bill; red – level 1: more expensive generation, with a greater

In matters related to SDG 7, which aims to ensure universal, reliable, modern and affordable access to energy, which has been partially met in its fundamental aspects, significant gaps persist. There is a need for more robust data to accurately assess the impact of high tariffs on low-income consumers, as well as the urgency of a deeper analysis of access to energy in isolated communities not connected to the National Interconnected System (SIN), especially in the North and Northeast regions of the country.

Launched in 2024 through a partnership with the Ministry of Science, Technology, and Innovation, the Solidarity Energy (*Energia Solidária*) Project stands out as an innovative initiative focused on the socio-economic inclusion of vulnerable communities. Within the energy transition context, the project, carried out in Brasília, in the Federal District, has the primary goal of democratising access to photovoltaic energy. It aims to raise public awareness of the environmental, economic and social benefits of this renewable source, train professionals for the renewable energy job market and donate and install solar panels for vulnerable families. Through educational strategies and sustainable technology implementation, the project contributes to the realisation of fundamental rights, such as the right to sustainable development, human dignity and the reduction of social inequalities, in line with constitutional principles and international commitments undertaken by the Brazilian State.[59] The project's primary directive is the technical training of at least 1,300 young people and adults, to be conducted in eleven learning laboratories strategically distributed across ten administrative regions of the Federal District. It also includes the installation of 1,000 solar panel kits in homes of low-income families, thereby promoting access to sustainable energy and social inclusion.[60]

increase in the bill; red – level 2: critical situation, with more expensive energy generation, resulting in a substantial increase in the electricity bill; Garcia, A. S. *et al.* 2024. Tariff flags and electricity consumption response in Brazil. *Utilities Policy,* 88. https://doi.org/10.1016/j.jup.2024.101751

[59] Ministry of Science, Technology and Innovation; Avante Brasil (OSCIP). 2024. Energia Solidária Project [Project funded through a fostering agreement – Process No. 01245.000351/2024-96]. Ministry of Science, Technology and Innovation. Available at: https://discricionarias.transferegov.sistema.gov.br/voluntarias/ConsultarProposta/ResultadoDaConsultaDeConvenioSelecionarConvenio.do?sequencialConvenio=966767&Usr=guest&Pwd=guest (Accessed:14 August 2025).

[60] Ministry of Science, Technology and Innovation; Avante Brasil (OSCIP). 2024. Work Plan – Energia Solidária Project [Project funded through a fostering agreement – Process No. 01245.000351/2024-96]. Available at: https://avantebrasil.ong.br/wp-content/up-

To anticipate potential resistance to solar energy adoption, the project involves local community leaders, integrating technical and scientific knowledge with well-tested photovoltaic technologies already commercialised in Brazil. The goal is to facilitate the technological implementation of the project's functional actions, promote active engagement of the target population and ensure the delivery of direct benefits to socio-economically vulnerable families, as well as the systematic dissemination of clean technologies within communities. The success of renewable energy in combating climate change depends on effective public engagement. Viewing the public as a barrier or ignorant is unhelpful; it's necessary to understand and mediate their concerns. This requires a comprehensive engagement approach, which faces both theoretical and practical challenges – from the way engagement is conceived to the difficulty of bringing together diverse interests, listening to them and genuinely integrating them into policy.[61]

Local cultural realities are respected, encouraging dialogue and collective participation. The knowledge of participants and project teams must be valued in a continuous process of informative feedback, collective knowledge construction and strengthening of social ties.[62] Project implementation is carried out in an integrated manner with participants, encouraging their active role and providing spaces for deliberation and decision-making aligned with the diverse needs and demands of the served communities.[63] The project is distributed across the following ten administrative regions in Brasília: Ceilândia, Taguatinga, Samambaia, Gama, Santa Maria 1, Santa Maria 2, Sobradinho, Sol Nascente, Planaltina, Plano Piloto and Lago Sul, through both fixed and mobile execution centres called Professional Activity centres.[64]

loads/2025/01/Plano_de_Trabalho_-_PRENERGIS_assinado.pdf (Accessed: 14 August 2025).

[61] MacArthur, J. 2016. Challenging public engagement: Participation, deliberation and power in renewable energy policy. *Journal of Environmental Studies and Sciences*, *6*(3), 631–640.

[62] Walker, G. *et al.* 2010. Trust and community: Exploring the meanings, contexts and dynamics of community renewable energy. *Energy policy*, 38(6): 2655-2663. https://doi.org/10.1016/j.enpol.2009.05.055

[63] Van Der Schoor, T., and Scholtens, B. 2015. Power to the people: Local community initiatives and the transition to sustainable energy. *Renewable and sustainable energy reviews*, *43*: 666–675. https://doi.org/10.1016/j.rser.2014.10.089

[64] Ministry of Science, Technology and Innovation; Avante Brasil (OSCIP). 2024. Work Plan – Energia Solidária Project [Project funded through a fostering agreement – Process No. 01245.000351/2024-96]. Available at: https://avantebrasil.ong.br/wp-content/up-

Methodologically, our analysis is also anchored in the framework of energy justice. This framework allows us to understand the Solidarity Energy project as an instrument of socio-urban regeneration. By articulating the principles of distributive, procedural, recognition and cosmopolitan justice, the project aligns with broader goals of sustainability and inclusion and presents itself as a concrete manifestation of energy justice. First, it ensures a fairer distribution of energy resources, contributing to the reduction of inequalities in energy access – an essential step toward distributive justice. Second, its collaborative and inclusive governance model guarantees active community participation in decision-making processes, promoting transparency and strengthening procedural justice. Furthermore, by deliberately incorporating historically marginalised and vulnerable groups, the project strengthens recognition justice, challenging the exclusionary structures that traditionally shape energy systems. Finally, by connecting local renewable energy production and consumption with global goals of climate justice and a just energy transition, it reinforces cosmopolitan justice, demonstrating that local solutions can contribute to broader transformations.

4.2.3 Gaps Between Energy Communities in Italy and Brazil

Brazil's lack of a legal framework for defining an energy community still represents a setback. An appropriate bill could define the rules for establishing a renewable energy community and also resolve the long-standing debate over the lack of a definition of energy poverty in the Brazilian context. In this context, beyond promoting the use of renewable energy, the Solidarity Energy Project also aims to prepare individuals for the job market by: training participants to safely and efficiently install photovoltaic energy systems through the provision of technical and practical knowledge regarding the installation and maintenance of their components, in compliance with local regulations and applicable safety standards;[65] fostering job and income generation, contributing to increased purchasing power

loads/2025/01/Plano_de_Trabalho_-_PRENERGIS_assinado.pdf (Accessed:14 August 2025).

[65] Sharma, S., and Sengar, N. 2022. Review of solar PV training manuals and development of survey based solar PV system training formats for beginners. *Solar Energy*, *241*, 72–84. https://doi.org/10.1016/j.solener.2022.05.055

among socio-economically vulnerable populations;[66] encouraging entrepreneurial skills development, with the goal of inspiring newly trained professionals to start businesses or work independently in the solar energy sector;[67] raising social awareness about the importance of photovoltaic energy, emphasising its role in mitigating the effects of climate change and promoting environmental sustainability. It's a broad perspective but still very diffuse, which could prevent good results from being consolidated over time if these advances don't become public policies. Encouraging public policies to leverage energy communities is a way to achieve greater legal certainty. Of course, the first step to this is a law on RECs in Brazil, which must be approved by the Brazilian legislature.[68]

On the other hand, both address vulnerable communities, but in different ways. Italy, following the European Union, is supporting the promotion of energy communities, while Brazil is encouraging the diffusion of energy in small groups with, at most, shared generation among members of the same family. Despite advances in urban sustainability, Brazil still faces significant challenges in reducing inequality and improving the environment, especially in interdependent metropolitan regions. In Brazil, disparities in SDG performance between regions – better in the South and worse in the North – reflect economic, historical, cultural, geographic and institutional factors. Cities adopt different solutions, and sharing experiences can drive improvements. Strengthening municipal education policies is recommended, given their positive impact on reducing poverty and inequality. Regions with better educational performance tend to be less unequal, except for the Federal District, where inequality persists due to the concentration and valorisation of jobs in the public sector, which is not matched by other sectors.[69]

[66] Falcone, P. M. 2023. Sustainable energy policies in developing countries: a review of challenges and opportunities. *Energies*, *16*(18). https://doi.org/10.3390/en16186682

[67] Mahajan, R., and Bandyopadhyay, K. R. 2021. Women entrepreneurship and sustainable development: select case studies from the sustainable energy sector. *Journal of Enterprising Communities: People and Places in the Global Economy*, *15*(1), 42–75. DOI: 10.1108/JEC-11-2020-0184

[68] Izam, N. S. M. N., *et al.* 2022. Sustainable development perspectives of solar energy technologies with focus on solar Photovoltaic—A review. *Energies*, *15*(8). https://doi.org/10.3390/en15082790

[69] Silva Martinelli, F., and Lindner, A. 2021. How far are the metropolitan areas in Brazil from achieving the sustainable development goals? An analysis based on SDG dashboards. *SN Social Sciences*, *1*(103). https://doi.org/10.1007/s43545-021-00131-8

Glaringly, a definition of energy poverty does not appear to be definitive for either Brazil or Italy, but while the Italian example cites project financing from private partners, with the support of a specific foundation to combat energy poverty, in Brazil, the role of the state, in the example cited (Solidarity Energy/*Energia Solidária*), was predominantly public because the resources invested in the project are public. Solidarity Energy seeks to make electricity more accessible through the Social Tariff program, benefiting approximately 10,000 low-income individuals registered in the Unified Registry for Federal Government Social Programs (Unified Registry) or who are beneficiaries of the Continuous Cash Benefit by offering discounts on electricity bills. In Brazil, the Unified Registry (*CadÚnico*) is the main tool used to identify and describe the socio-economic status of low-income families across the country. Enrolment in the Unified Registry is essential for families to qualify for access to various social programs, such as Family Allowance, the Social Tariff for Electricity and Continuous Cash Benefit, among others. However, it is important to highlight that simply registering in the Unified Registry does not guarantee automatic inclusion in these programs, as each program has its own specific criteria and eligibility requirements.[70]

While in Italy energy poverty is monitored by the Italian Observatory on Energy Poverty (OIPE) with a statistical focus, in Brazil there is a paradox between subsidies and consumption that reveals access gaps. The new Social Tariff for Electricity covers elderly individuals and people with disabilities who receive Continuous Cash Benefit, indigenous and quilombola families registered in CadÚnico and those served by isolated systems powered by solar panels and batteries. Currently, families enrolled in the Unified Registry who have a monthly income of up to half the minimum wage per person and who consume up to 80 kilowatt-hours per month are exempt from paying for the electricity they consume. There is no single legal and official definition of energy poverty in Brazil, but considering that it represents the lack of access to modern and reliable energy services by individuals or families, it can be understood that the government recognises that those benefiting from the new social electricity tariff are in energy poverty.[71] In addition to the traditional definition of

[70] Sarlet, I. W. 2022. *Social security in Brazil: Redefining minimum income support (No. 7)*. Social Law Report. Max Planck Institute for Social Law and Social Policy.

[71] Delgado, M., and Rufin, C. 2025. Energy transition policies in the Global North and the Global South: Comparing the US and Brazil. *Social Science Research Network*. https://

energy poverty, which is defined as when a family cannot access adequate energy services for a 10 percent income, the current definition of energy poverty must take into account information technology and appliances essential for everyday life.[72] Consumer vulnerability and energy poverty, as synonyms, are not easily identified. Especially when it comes to energy poverty, there is no legal definition for the topic in Italy.[73]

What Solidarity Energy aims to achieve is to allow the installation of 1,000 photovoltaic panel kits in homes belonging to low-income families, thereby ensuring access to sustainable energy and promoting social inclusion.

Italy could adopt from Brazil a better, clearer and more operational definition of energy poverty, since in Brazil, energy poverty is associated with objective criteria such as income, consumption and registration with the Unified Registry (*CadÚnico*). This facilitates policy targeting (e.g., Social Tariff, programs like Solidarity Energy). Furthermore, there is greater integration between energy and social inclusion, as the Solidarity Energy Project connects photovoltaic generation to professional training, income and inclusion, training new professionals for the job market. This is about creating social and economic opportunities. On the other hand, Brazil could adopt the stable regulatory framework that Italy has, since Brazil still lacks a specific legal framework for RECs. Although legislation on energy communities continues to be debated, something is better than nothing. Progress on this point would bring stability and scale. Encourage greater use of cooperative and solidarity-based redistribution models, as Le Vele REC guarantees that 100% of energy-sharing revenues return to the community in social services and support.

4.3 Conclusion

Today's cities stand at a critical crossroads. While they face the pressing challenges of climate change, social inequality and energy vulnerability, they also remain vital laboratories of innovation and inclusion. Urban

dx.doi.org/10.2139/ssrn.5117055

[72] Garetto, R. 2021. Overcoming energy poverty through becoming a prosumer? In L. Ruggeri (Ed.), *Needs and Barriers of Prosumerism in the Energy Transition Era* (pp. 43–55). Dykinson.

[73] Giobbi, M. 2021. *Il consumatore energetico nel prisma del nuovo quadro regolatorio italo-eurounitario* (pp. 136–137). Napoli: Edizioni Scientifiche Italiane.

regeneration emerges as a necessary response to uncontrolled urban growth, but it becomes truly effective only when conceived as an integrated process that simultaneously addresses environmental, social and economic dimensions. This chapter aimed to highlight the role of RECs and, broadly, of renewable energy projects as key enablers of socio-urban regeneration. The case of *Le Vele REC* in Roma illustrates how energy communities can lower costs, strengthen community ties and improve the living conditions of vulnerable groups. Similarly, the *Energia Solidária* initiative in Brasília demonstrates that collective solar generation can reduce poverty, empower communities and promote the dissemination of renewable technologies, even in the absence of a dedicated legal framework. Despite their different contexts, both experiences show how renewable energy projects can transform disadvantaged areas into productive, community-orientated infrastructures, restoring value and identity, stimulating local economies and fostering resilient neighbourhoods capable of tackling social and climate challenges.

In light of these examples, cities – when coupled with energy communities and other renewable energy projects – become focal points of new *ecologies of proximity*,[74] where environmental, social and economic dimensions converge within a regenerative approach. The principles of energy justice provide the guiding framework for this trajectory. Prioritising energy justice in urban regeneration enables the integration of technical infrastructures with human rights, sustainability and social inclusion imperatives. Energy communities therefore emerge as both ecological devices and social infrastructures, capable of mitigating inequalities, strengthening democratic participation and advancing the UN SDGs. They pave the way for a transformative vision of future cities, where tackling energy poverty aligns with the broader ambition of creating beautiful, sustainable and inclusive urban environments.[75]

[74] D'Onofrio, R., and Trusiani, E. 2022. The Future of the City in the Name of Proximity: A New Perspective for the Urban Regeneration of Council Housing Suburbs in Italy after the Pandemic. *Sustainability*, *14*(3). https://doi.org/10.3390/su14031252

[75] European Commission. 2025. *New European Bauhaus: beautiful, sustainable, together*. Available at: https://new-european-bauhaus.europa.eu/index_en (Accessed: 21 August 2025).

PART III

Reuse: Maximise Resource Efficiency and Sustainability by Managing Consumption

CHAPTER 5

Policy Trade-offs in a Just Transition Framework: Examining the Fast Fashion Crisis in Bangladesh Through a Degrowth Lens

Nils Grolimund, Shuddha Srimoyee Das, Laura Peña, and Silvia Ciacchi

Abstract Sustainability, as a holistic and multifaceted concept, encompasses interconnected human rights and environmental dimensions, exemplified by the fast fashion industry's reliance on exploitative practices such as child labour and systemic worker rights violations, alongside severe environmental degradation through pollution, toxic emissions and ecosystem damage. This industry's unsustainable model is further characterised by a geographical disparity: production and waste management burdens disproportionately fall on

N. Grolimund
University of Lausanne, Lausanne, Switzerland
e-mail: NilsMartin.Grolimund@unil.ch

S. S. Das
United Nations University-Maastricht Economic and Social Research Institute on Innovation and Technology (UNU-MERIT), Maastricht, The Netherlands

L. Peña
Centre for Environment, Economy and Energy, Brussels School of Governance, Vrije Universiteit Brussels, Brussels, Belgium

S. Ciacchi
Erasmus University Rotterdam, Rotterdam, The Netherlands

A. Phillips, L. Kaschny (eds.), *Global Perspectives on the Circular Economy in the Just Transition*, Just Transitions,
https://doi.org/10.1007/978-3-032-18342-2_5

the Global South, where labour exploitation and inadequate infrastructure exacerbate ecological crises, while consumption-driven demand originates predominantly in the Global North, perpetuating a cycle of environmental injustice and social inequity. Sustainability policy initiatives are often pushed by social mobilisation in the wake of public scandals and disasters. Because of this, they tend to be reactive and, consequently, limited to the scope of the catalyst incident. This dynamic is exemplified by the Rana Plaza collapse and the following policy action. The Rana Plaza incident refers to the 2013 collapse of a garment factory building in Bangladesh, which killed over 1,100 workers and injured thousands more. The tragedy exposed the unsafe working conditions and lack of oversight in the global fashion supply chain, driven by the demand for cheap, fast production. It became a symbol of the urgent need for ethical labour practices and corporate accountability. In this chapter, we use Rana Plaza as a case study to show how such disasters highlight the limitations of narrowly focused reforms and the need for holistic sustainability approaches that address not just labour rights but also environmental degradation, overconsumption and global economic inequalities. Rana Plaza was instrumental in mobilising social pressure for regulatory change and in driving the implementation of much-needed policy initiatives. However, because of the nature of the disaster, policies tended to focus on workers' rights and social security. This chapter shines a light on the limits of sectoral reactive policy in the face of holistic sustainability issues and draws attention to the fact that narrowly scoped policy interventions miss the larger picture at best and generate unexpected knock-on effects at worst. For example, although workers' rights are crucial to the fast fashion issue, true holistic sustainability cannot be achieved without changes in consumption and reductions in the levels of production. Indeed, current production levels exacerbate environmental externalities as well as mutually reinforce the dynamics of social injustice in producer countries. Building a Just Transition framework on degrowth principles could tackle these issues. Simple worker-orientated policy initiatives miss the environmental picture, which would be better addressed through degrowth policy and reduced consumption in the global North. However, simply focusing on reductions without targeted institutional reforms in producer countries would produce grave socio-economic effects. Holistic problems require holistic solutions. Degrowth policies in the global North should go hand in hand with strengthening institutions, social security and economic reform in the Global South and producer countries. One cannot happen without the other.

The question is: can global leaders achieve the level of coordination required for holistic and synergetic policy action?

Keywords Fast Fashion • Degrowth • Rana Plaza • Extended Producer Responsibility • Circularity • Reactive Policy • Holistic Policy

5.1 Introduction

Sustainability is a holistic, multifaceted and interconnected concept that ranges from human rights to environmental preservation. These interconnected and multifaceted dimensions of sustainability can be highlighted well by examining the fashion industry and the fast fashion sector more concretely, whose reliance on exploitative practices such as child labour and systemic workers' rights violations is compounded by severe environmental degradation through pollution, toxic waste and emissions and ecosystem damage by its production processes.[1] The industry's unsustainable practices are further characterised by geographical disparities: production and waste management burdens disproportionately fall on the Global South, where labour exploitation and inadequate infrastructure exacerbate ecological crises, while consumption-driven demand originates predominantly in the Global North, perpetuating a cycle of environmental injustice and social inequity.[2]

Over the last decades, public awareness about issues and challenges relating to sustainability has increased dramatically. This awareness has brought with it scrutiny, social mobilisation and increasing public pressure for regulatory action.[3] This social pressure-driven mode of policy change results in 'reactive policies' whose scope relates only to the issues exposed by the inciting event.[4] Due to this dynamic, issues related to sustainability

[1] Suraci, O. 2021. The Best-Dressed Polluter – Regulation and Sustainability in the Fashion Industry. *Hastings Environmental Law Journal, 27*(2): 225; Centobelli, Piera. *et al.* 2022. Slowing the fast fashion industry: An all-round perspective. *Current Opinion in Green and Sustainable Chemistry, 38* https://doi.org/10.1016/j.cogsc.2022.100684

[2] Suraci, O. 2021. The Best-Dressed Polluter – Regulation and Sustainability in the Fashion Industry. *Hastings Environmental Law Journal, 27*(2): 225; Bhambra, G. K. 2022. A Decolonial Project for Europe. *JCMS: Journal of Common Market Studies, 60*(2): 229–244. https://doi.org/10.1111/jcms.13310

[3] UN Global Compact. Rana Plaza Factory: Global Outcry Triggers Industry Action. Business & Human Rights Navigator. Available at: https://bhr-navigator.unglobalcompact.org/case_studies/global-outcry-triggers-industry-action/ (Accessed: 20 August 2025).

[4] Choudhury, M and Haque, C.M. 2024. Disaster management policy changes in Bangladesh: Drivers and factors of a shift from reactive to proactive approach. *Environmental Policy and Governance, 34*(5):445–462. https://doi.org/10.1002/eet.2094

are tackled through a piecemeal approach, making it difficult to achieve a just transition that reconciles both environmental and social aspects. This chapter aims to highlight this apparent disconnect between the global and holistic nature of sustainability issues and the sectoral and targeted scope of reactive policies. Bangladesh is used as a case study for this problem, as it is one of the world's leading producers of Ready-Made Garments (RMG).[5] Consequently, both the socio-economic and ecological systems of Bangladesh are deeply affected by fast fashion's negative externalities.[6] Furthermore, Bangladesh was also the stage for one of the most (in)famous examples of crisis-driven social mobilisation in recent years, namely, the Rana Plaza collapse of 2013.

The Rana Plaza collapse occurred in 2013 and involved the catastrophic collapse of an eight-storey commercial building, killing over 1,100 workers and injuring thousands more.[7] The disaster put the dire working conditions in the fashion supply chain onto the world stage, revealing the systemic neglect of safety standards in the pursuit of low-cost, high-volume production.[8] As a result, the Rana Plaza collapse became a turning point in conversations connecting fashion and sustainability, highlighting the urgent need for ethical and humane labour practices as well as corporate

[5] RMG is understood here as garments that are mass-produced and are ready to wear at sale, thereby differing from tailor-made clothing. Available at: https://www.garmentexporthouse.com/2021/09/rmg-meaning-in-garment-industry.html (Accessed: 20 August 2025). For the relevance of the RGM industry for Bangladesh see: Sharif, I.S. and Das, D. 2022. Bangladesh's Emergence as a Ready-Made Garment Export Leader: An Examination of the Competitive Advantages of the Garment Industry. *International Journal of Global Business and Competitiveness, 17*: 162–174. https://doi.org/10.1007/s42943-022-00049-9; World Trade Statistics. 2025. Available at: https://www.wto.org/english/res_e/statis_e/world_trade_statistics_e.htm (Accessed: 20 August 2025).

[6] Alam, H. *et al.* 2024. International Circular Economy Policy Trends - Implications for Bangladesh's Circular Garment and Textile Policy Development. Available at: https://www.switchtocircular.eu/sites/default/files/publications/International%20circular%20economy%20policy%20trends_implications%20for%20Bangladesh_pub_web.pdf (Accessed:20 August 2025); Swedwatch. 2024. Paying the price for fashion: securing a living wage for Bangladesh's garment workers. Available at: https://swedwatch.org/wp-content/uploads/2024/11/briefsecuring-living-wagesnov-201124.pdf (Accessed: 20 August 2025).

[7] ILO. 2023. The Rana Plaza disaster ten years on: What has changed? Available at: https://webapps.ilo.org/infostories/en-GB/Stories/Country-Focus/rana-plaza#intro (Accessed: 20 August 2025).

[8] Barua, D. 2025. The Transfer from Fast Fashion to Slow Fashion: Challenges and Opportunities for Bangladesh's RMG Sector. *International Journal on Science and Technology: 16*(3).

accountability in global garment manufacturing.[9] In this chapter, the Rana Plaza case is used to highlight the need for more holistic approaches to sustainability that account for the deeply interwoven nature of social justice, environmental degradation and economic dependence. While the collapse brought important reforms, the responses remained limited,[10] and, as this chapter examines, too narrowly focused on workers' rights and worker safety, overlooking broader systemic issues such as overconsumption in the Global North or environmental harm. Indeed, the exploitation of Bangladeshi workers is a symptom of larger issues of overconsumption and overproduction that shape the global economy.[11] Rana Plaza exemplifies how reactive, sectoral policies fail to tackle the root causes of unsustainability well.[12] Upon this observation, this chapter argues that a more holistic approach to sustainability would require integrated solutions that consider the full lifecycle, from production and consumption to disposal and the transnational power asymmetries that shape them. Additionally, as discussed throughout, a truly holistic approach must integrate labour rights, ecological limits and economic restructuring across both producer and consumer contexts.

A common approach to tackle environmental degradation and social issues broadly is the Just Transition framework. Originally, it emerged in the U.S. labour movement of the 1980s as an avenue to protect workers

[9] On the pivotal role of Rana Plaza see *inter alia*: ILO. n.d. A turning point for occupational safety and health in supply chains. Available at: https://vzf.ilo.org/Photo%20Exhibition/a-turning-point-for-occupational-safety-and-health-in-supply-chains/#:~:text=After%20the%20disaster%2C%20the%20global,of%20profit%20and%20economic%20growth (Accessed: 20 August 2025); industriAll Europe. 2023. Rana Plaza tragedy: a turning point for a safer garment industry. Available at: https://news.industriall-europe.eu/Article/893 (Accessed: 20 June 2025).

[10] IHRB. 2023. Rana Plaza 10 Years On: Lessons for Human Rights and Business. Available at: https://ihrb-org.files.svdcdn.com/production/assets/uploads/briefings/rana-plaza-lessons-for-human-rights-and-business.pdf (Accessed: 20 August 2025).

[11] D'Alisa, G. *et al.* 2015. Degrowth: A Vocabulary for a New Era. Routledge; Fletcher, Kate, and Mathilda Tham. 201; Earth Logic: Fashion Action Research Plan. Available at: https://katefletcher.com/wp-content/uploads/2019/10/Earth-Logic-plan-FINAL.pdf (Accessed: 20 August 2025).

[12] Lohmeyer, N. and Schüßler, E. 2019. Rana Plaza as a Threat to the Fast Fashion Model? An Analysis of Institutional Responses to the Disaster in Germany. In: Becker-Leifhold, and M. Heuer (Eds.) *Eco Friendly and Fair: Fast Fashion and Consumer Behavior*. 3–14. London and New York: Routledge.

and communities from industrial change[13] and it has since broadened to encompass multiple dimensions of justice, namely procedural, restorative and recognition-based concerns.[14] International bodies, such as the International Labour Organization (ILO), frame it as a process of greening the economy to ensure decent work and leave no one behind. Academically, interpretations of the Just Transition framework vary from reformist approaches focused on 'green growth'[15] to radical calls for systemic change.[16] This plurality of understanding means that many people can work with it; however, it also means that it is applied unevenly: some implementations prioritise technical or sectoral fixes, while others aim to challenge the political-economic structures underpinning sustainability. In this chapter, our aim is to highlight how, without explicitly problematising the structural drivers, such as overconsumption and overproduction, any Just Transition framework is at risk of being reduced to a patchwork of well-intentioned but fragmented measures. To tackle this problem, degrowth can serve as the underlying foundation of a Just Transition framework that has planetary boundaries at its core.

This chapter is structured as follows. Section 2 introduces the Rana Plaza disaster in Bangladesh and outlines how the fashion industry contributes to interconnected human rights violations and environmental degradation in the country. In Section 3, we focus on the ensuing policy responses. Section 4 shifts the focus to programmes and policy frameworks that originate from the Global North. It outlines how existing initiatives have improved sustainable practices throughout the supply chain but also which limitations and risks exist by not addressing the political-economic root causes of unsustainability, namely overconsumption and overproduction. Section 5 introduces degrowth as the normative and political

[13] Newell, P. and Mulvaney, D. 2013. The political economy of the 'just transition.' *The Geographical Journal, 179(*2): 132–140. https://doi.org/10.1111/geoj.12008

[14] McCauley, D. and Heffron, R. 2018. Just transition: Integrating climate, energy and environmental justice. *Energy Policy, 119*: 1–7. https://doi.org/10.1016/j.enpol.2018.04.014; Heffron, R. 2021. *Achieving a Just Transition to a Low-Carbon Economy*. Palgrave Macmillan Cham.

[15] Stevis, D. and Felli, R. 2020. Planetary just transition? How inclusive and how just? *Earth System Governance, 6*. https://doi.org/10.1016/j.esg.2020.100065

[16] Cigna, L. *et al.* 2023. Varieties of Just Transition? Eco-Social Policy Approaches at the International Level. *Social Policy and Society, 22*(4): 730–746. https://doi.org/10.1017/S1474746423000192

approach that can give existing just transition initiatives an improved whole-system perspective. In the conclusion, the importance of integrated and multilevel global sustainability policies and holistic approaches to a just transition are elaborated.

5.2 The Problem of Unsustainable Fashion: The Rana Plaza Disaster

The fast fashion business model is quite easy to understand: follow the latest trends and produce rapidly at low costs.[17] Due to the high frequency of new items, consumers often feel like these items are disposable and therefore serve their impulse to buy more frequently.[18] To satisfy the high demand for these products, companies outsource their production to countries where they can exploit the lax legislative system and poor working conditions. As one of the leading producers of RMG, Bangladesh is situated in a peculiar position. While 4 million workers are employed in the garment industry, contributing to 80% of the country's exports, the production processes are resource intensive and come with a high amount of waste and emissions.[19] While some regulations have been agreed upon to support the workers' socio-economic position, only 23% of companies follow through on the agreement to support childcare facilities, for example.[20] The Bangladesh Labour Act of 2006[21] stipulates that companies with 40 or more female workers must provide childcare support; however, most do not. Consequently, many women are left with no choice but to

[17] For greater context see: Kelleher, Dylan. 2025. Fast Fashion. Brittanica, 2025. Available at: https://www.britannica.com/art/fast-fashion. (Accessed: 20 August 2025).

[18] Centobelli, P. *et al.* 2022. Slowing the fast fashion industry: An all-round perspective. *Current Opinion in Green and Sustainable Chemistry, 38.* https://doi.org/10.1016/j.cogsc.2022.100684

[19] Alam, H. *et al.* 2024. International Circular Economy Policy Trends - Implications for Bangladesh's Circular Garment and Textile Policy Development. Available at: https://www.switchtocircular.eu/sites/default/files/publications/International%20circular%20economy%20policy%20trends_implications%20for%20Bangladesh_pub_web.pdf

[20] International Finance Corporation. 2019. Tackling Childcare: The Business Benefits and Challenges of Employer-Supported Childcare in Bangladesh. Available at: https://www.ifc.org/content/dam/ifc/doc/mgrt/final-ifc-tackling-childcare-bangladesh-110619.pdf (Accessed: 20 August 2025).

[21] Bangladesh Labour Act, 2006 (Act No. XLII of 2006), enacted on 11 October 2006.

quit their jobs to provide care to their children, leaving important income behind.[22]

Additionally, irresponsible purchasing by international buyers has also contributed to low wages and poor living conditions. While factory management often feels that there is a lack of transparency in the buyers' pricing, this power imbalance is ultimately ignored in fear of losing their business.[23] This power imbalance is also clearly visible in the aftermath of the Tazreen fire, which claimed at least 112 lives, yet the process to ensure compensation for victims was extremely tedious.[24]

A turning point for policies finally came in response to the Rana Plaza collapse, which could have been prevented. Cracks in the building had already been observed the previous day; however, the garment factory section was kept operating while businesses on the lower levels chose to close after a warning was issued.[25] The RMG workers in the building reluctantly came to work, as they were threatened with job loss from management.[26] Tragically, their lives were changed forever, and after months of rescue operations and years of demanding justice and compensation, it is still necessary to make progress. This disaster brutally exposed how little the lives of RMG workers matter in the supply chain and the injustices they are facing in Bangladesh.

In response to the collapse, North American companies started The Alliance for Bangladesh Worker Safety (the Alliance),[27] while the European companies prompted The Bangladesh Accord on Fire and Building Safety

[22] Swedwatch. 2024. Paying the price for fashion: securing a living wage for Bangladesh's garment workers. Available at: https://swedwatch.org/wp-content/uploads/2024/11/briefsecuring-living-wagesnov-201124.pdf (Accessed: 20 August 2025).

[23] Ibid.

[24] Amnesty International. 2024. Garment workers must receive rights-based compensation and justice immediately. Available at: https://www.amnesty.org/en/latest/news/2024/05/bangladesh-garment-workers-must-receive-rights-based-compensation-and-justice-immediately/ (Accessed: 20 August 2025).

[25] Akhter, S. 2014. Endless Misery of Nimble Fingers: The Rana Plaza Disaster. *AJWS, 20*(1): 137–147. https://doi.org/10.1080/12259276.2014.11666176; Clean Clothes Campaign. Rana Plaza. Clean Clothes Campaign. Available at: https://cleanclothes.org/campaigns/past/rana-plaza (Accessed: 20 August 2025).

[26] Clean Clothes Campaign. Rana Plaza. Clean Clothes Campaign. Available at: https://cleanclothes.org/campaigns/past/rana-plaza (Accessed: 20 August 2025).

[27] The Alliance for Bangladesh Worker Safety. n.d. The Alliance for Bangladesh Worker Safety. https://bangladeshworksafety.org (Accessed: 20 August 2025).

Table 5.1 Key differences and similarities between the Bangladesh Accord on Fire and The Accord- Building Safety, Adopted from Rahman and Rahman, 2020.

Differences	*Reactive Measure*	
	The Accord	*The Alliance*
Legally Binding	✔	✖
Fire, Electrical and Building Safety Focused	✔	✔
Worker Training and Empowerment	✔	✔
Extent of Cooperation	Transnational co-determination	Industry self-regulation
Governance System	Labour-driven	Corporate-driven
Responsibility in case of factory closure	Workers compensated by Employers	Compensation costs shared by employers and buyers

Rahman, S. and Rahman, K.M.. 2020. Multi-actor Initiatives after Rana Plaza: Factory Managers' Views. *Development and Change*, *51*(5): 1331–1359. https://doi.org/10.1111/dech.12572

(the Accord).[28] Whilst the Alliance ceased operations in 2018, the Accord is still ongoing. Table 5.1 shows the key differences and similarities between the policy approaches. They share a focus on fire, electrical and building safety, as well as worker training and empowerment. Their primary differences are in terms of bindingness, governance structures, extent of cooperation and the allocation of responsibility for compensation in the event of factory closures.

The aftermath of the disaster showcases that, if the 'right' types of pressures exist, beneficial policies can be put in place, overriding any lack of political will. In other words, the Rana Plaza disaster 'had to' happen to produce the necessary pressures for beneficial policies to be taken seriously in Bangladesh, in addition to creating pressure on the factories to ensure better working conditions.

[28] The Accord on Fire and Building Safety in Bangladesh. n.d. The Accord on Fire and Building Safety in Bangladesh. https://bangladeshaccord.org (Accessed: 20 August 2025); See generally: IHRB. 2023. Rana Plaza 10 Years On: Lessons for Human Rights and Business. Available at: https://ihrb-org.files.svdcdn.com/production/assets/uploads/briefings/rana-plaza-lessons-for-human-rights-and-business.pdf (Accessed: 20 August 2025).

5.3 Reactive Policy Initiatives Post-Rana Plaza: A Focus on Workers' Rights

The Rana Plaza collapse achieved a level of publicity, indignation and reputational whiplash the Tazreen factory fire did not. The disaster exposed the widespread negligence and the unsafe working conditions and opened the curtain to expose the exploitative nature of the fast fashion supply chain. The ensuing international outrage was bolstered by the involvement of several Western brands such as H&M, Zara, Benetton, Primark, Walmart and more.[29] The consequence was a cry for safety reforms and corporate accountability.[30]

The disaster led to institutional action in the form of the Alliance, the Accord and the Bangladesh Sustainability Compact,[31] an international political agreement between the Government of Bangladesh, the European Commission, the ILO, the United States and Canada on the same themes. However, these reforms remained focused on the most visible and politically actionable causes of the collapse, namely worker safety and labour rights. This narrow focus fails to account for fast fashion's structural causes of the disaster. Lohmeyer and Schüßler refer to this phenomenon as 'production-orientated initiatives' versus what could have been 'consumption-orientated initiatives.'[32] The scope and focus of these institutional responses are especially poignant considering their effectiveness. As the ILO noted in its ten-year review,[33] the improvements in working

[29] Guinebault, M. 2023. Rana Plaza: ten years later, has the fashion industry learned its lesson? *Fashion Network*. Available at https://ww.fashionnetwork.com/news/Rana-plaza-ten-years-later-has-the-fashion-industry-learned-its-lesson-,1506162.html (Accessed: 20 August 2025).

[30] UN Global Compact. Rana Plaza Factory: Global Outcry Triggers Industry Action. Business & Human Rights Navigator. Available at: https://bhr-navigator.unglobalcompact.org/case_studies/global-outcry-triggers-industry-action (Accessed: 20 August 2025).

[31] ILO. n.d. The Sustainability Compact for the Bangladesh Ready-Made Garment Sector. Available at: https://mhssn.igc.org/EU%20Sustain%20Compact%20-%20BGD%20-%20Jan%202019%20.pdf. (Accessed: 20 August 2025).

[32] Lohmeyer, N. and Schüßler, E. 2019. Rana Plaza as a Threat to the Fast Fashion Model? An Analysis of Institutional Responses to the Disaster in Germany. In: C. Becker-Leifhold, and M. Heuer (Eds.) *Eco Friendly and Fair: Fast Fashion and Consumer Behavior*, 3–14. London and New York: Routledge.

[33] ILO. 2023. The Rana Plaza disaster ten years on: What has changed? Available at:https://webapps.ilo.org/infostories/en-GB/Stories/Country-Focus/rana-plaza#intro (Accessed: 20 August 2025).

conditions achieved through factory inspections and remediation efforts have largely remained confined to export-orientated facilities.[34] Numerous subcontracted factories remain outside regulatory frameworks to date and Bangladesh's domestic labour protections are still marked by enforcement gaps, inadequate social safety nets and institutional fragmentation.[35]

In their 2024 paper, Choudhury and Haque analyse the Bangladeshi (environmental) disaster management policy approaches.[36] Their analysis builds on the established literature on the drivers of policy change, which highlights how such changes are often triggered by shocks from natural disasters,[37] also referred to as 'focusing events.'[38] Choudhury and Haque specifically contrast this type of 'reactive' policy change, which relies on a window of opportunity for policy action generated by the focusing event, with a proactive approach, which is geared towards the application of available knowledge and resources in favour of *ex ante* risk prevention.[39]

The crisis-driven modality of policy intervention can be seen in the institutional responses to the Rana Plaza disaster and the ensuing attempts for fast fashion reforms in Bangladesh. Although this was not a 'natural

[34] IHRB. 2023. Rana Plaza 10 Years On: Lessons for Human Rights and Business. https://www.ihrb.org/resources/briefing-rana-plaza-lessons-for-human-rights-business (Accessed: 20 August 2025); Saxena, S. B., and Baumann-Pauly, D. 2019. Off the radar: Subcontracting in Bangladesh's RMG industry. In: Sanchita B. Saxena (Eds). *Labor, Global Supply Chains, and the Garment Industry in South Asia*, 45–61. London: Routledge.

[35] ILO. 2023. The Rana Plaza disaster ten years on: What has changed? Available at: https://webapps.ilo.org/infostories/en-GB/Stories/Country-Focus/rana-plaza#intro (Accessed: 20 August 2025).

[36] Choudhury, M. and Haque, C.M. 2024. Disaster management policy changes in Bangladesh: Drivers and factors of a shift from reactive to proactive approach. *Environmental Policy and Governance, 34*(5): 445–462. https://doi.org/10.1002/eet.2094

[37] Albright, E. A. 2011. Policy Change and Learning in Response to Extreme Flood Events in Hungary: An Advocacy Coalition Approach. *Policy Studies Journal, 39*(3): 485–511.Available at: https://doi.org/10.1111/j.1541-0072.2011.00418.x; Haque, C. E. *et al.* 2019. 'Events and failures are our only means for making policy changes': learning in disaster and emergency management policies in Manitoba, Canada. *Natural Hazards, 98*(1): 137–162. https://doi.org/10.1007/s11069-018-3485-7

[38] Birkland, T. A. 1998. Focusing Events, Mobilization, and Agenda Setting. *Journal of Public Policy, 18*(1): 53–74. Available at: https://www.jstor.org/stable/4007601; Birkland, Thomas A. 2009. Disasters, Lessons Learned, and Fantasy Documents. *Journal of Contingencies and Crisis Management, 17*(3): 146–156. https://doi.org/10.1111/j.1468-5973.2009.00575.x

[39] DeLeo, R. A. 2017. Anticipatory policymaking in global venues: Policy change, adaptation, and the UNFCCC. *Futures, 92*: 39–47. https://doi.org/10.1016/j.futures.2016.09.001

disaster' by any means, the shocking impact of the collapse and the consequent mobilisation for both legal and policy change followed the same patterns as the reactive policy change described above. This dynamic is further illustrated by the absence of meaningful systemic reforms in response to the preceding Tazreen fire, underscoring how reforms remain contingent on the magnitude of public outrage rather than institutional foresight.

The results of the post-Rana Plaza initiatives exemplify the limits of reactive policies. While the collapse drew unprecedented international attention and spurred avenues for positive progress for worker protection, these interventions were mainly driven by external pressure and framed around immediate fixes rather than structural transformation. These responses failed to address the tensions and systemic failures that underlie the fast fashion system, namely overproduction, overconsumption and the global economic asymmetries that underpin the fast fashion model. Policymakers have since realised that it is necessary to expand the policy frameworks if a transition to sustainable business practices is to be achieved. To enable a just transition in the fashion sector, the scope of improvements needs to incorporate not only labour rights and safety but also the inclusion of ecological, social and economic systems. These holistic policy approaches have come into existence over the last years, how 'holistic' and successful they really are is the focus of the next section.

5.4 Holistic Approaches and the Dangers of Unilateralism

The limitations of reactive, sectoral policymaking in the fast fashion industry, as illustrated by the consequences of the Rana Plaza collapse, have ignited a shift towards more 'holistic' policy frameworks, especially stemming from the Global North. These approaches, such as lifecycle assessments (LCA), extended producer responsibility (EPR) and circular economy strategies, aim to address sustainability in an integrated manner, recognising the interconnectedness of social, environmental and economic dimensions.[40]

[40] European Commission. 2022. First Circular Economy Package including Regulation proposal on Sustainable Products and Sustainable Textiles Strategy. Available at: https://ec.europa.eu/newsroom/env/items/742027/en (Accessed: 20 August 2025; Mallick, P. K. *et al.* 2024. Designing and operationalising extended producer responsibility under the EU Green Deal. *Environmental Challenges, 16* https://doi.org/10.1016/j.envc.2024.100977

In principle, this shift represents progress. While previous initiatives narrowly targeted workplace safety or labour standards in response to a crisis, these holistic frameworks attempt to consider the full product life-cycle from design and production to disposal.[41] For example, the EU's Strategy for Sustainable and Circular Textiles promotes design for durability, supply chain transparency and the right to repair.[42] Similarly, the Ellen MacArthur Foundation's circular economy model encourages reuse and recycling in an attempt to decouple economic activity from resource extraction.[43] However, these frameworks often fall short of addressing the structural drivers that make the fast fashion industry so harmful, particularly for countries like Bangladesh.[44] One major shortcoming is the tendency to still treat social conditions and environmental outcomes as distinct problems.[45] Post-Rana Plaza reforms improved factory safety but did not challenge the economic model that demands cheap, disposable garments at scale.[46] Likewise, circular economy strategies that prioritise downstream recycling without reducing overall production risk reproduce the same linear, extractive dynamics under a new banner.[47]

[41] Eeckhout, Lieven. 2023. Towards sustainable computer architecture: A holistic approach. In: Duranton, M. *et al.* (Eds) *HiPEAC Vision 2023*; Marti, E. *et al.* 2024. The Impact of Sustainable Investing: A Multidisciplinary Review. *Journal of Management Studies, 61*(5): 2181–2211. https://doi.org/10.1111/joms.12957

[42] European Commission. 2022. First Circular Economy Package including Regulation proposal on Sustainable Products and Sustainable Textiles Strategy. Availabe at: https://ec.europa.eu/newsroom/env/items/742027/en (Accessed: 20 August 2025).

[43] Ellen MacArthur Foundation. 2017. A new textiles economy: Redesigning fashion's future. Available at: https://content.ellenmacarthurfoundation.org/m/6d5071bb8a5f05a2/original/A-New-Textiles-Economy-Redesigning-fashions-future.pdf (Accessed: 20 August 2025).

[44] Swedwatch. 2024. Paying the price for fashion: securing a living wage for Bangladesh's garment workers. Available at: https://swedwatch.org/wp-content/uploads/2024/11/briefsecuring-living-wagesnov-201124.pdf (Accessed: 20 August 2025).

[45] Lifset, R. *et al.* 2023. Restoring the incentives for eco-design in extended producer responsibility: The challenges for eco-modulation. *Waste Management, 168*: 189–201. https://doi.org/10.1016/j.wasman.2023.05.033

[46] Fletcher, K. and Tham, M. 2019. Earth Logic: Fashion Action Research Plan. Available at: https://katefletcher.com/wp-content/uploads/2019/10/Earth-Logic-plan-FINAL.pdf (Accessed: 20 August 2025).

[47] Suraci, O. 2021. The Best-Dressed Polluter – Regulation and Sustainability in the Fashion Industry. *Hastings Environmental Law Journal* 27(2): 225; Bhambra, G. K. 2022. A Decolonial Project for Europe. *JCMS: Journal of Common Market Studies, 60*(2): 229–244. https://doi.org/10.1111/jcms.13310

This highlights the first danger of the holistic frameworks. Lifecycle thinking that remains overly technical can miss the underlying political economy of overproduction and overconsumption.[48] As discussed before, the Global North's consumption habits are one major driver for low-cost production, pushing factories in Bangladesh to cut corners and keep wages low.[49] Without policies to tackle the demand, even the best design standards cannot ensure fair working conditions or meaningful ecological limits. A second danger arises in relation to unilateralism. The sustainability frameworks designed and imposed by institutions in the Global North without adequate input from producer countries risk reinforcing the same power asymmetries that are at the foundation of the fast fashion system.[50] While international due diligence initiatives like the EU's Corporate Sustainability Due Diligence Directive aim to hold companies accountable for harm throughout their supply chain; they often reflect Northern priorities, goals and norms.[51]

An important example of this tension is the EU's approach to EPR and the proposed Eco-design for Sustainable Products Regulation (ESPR).[52] On paper, EPR is meant to make producers financially and operationally responsible for the full life cycle impacts of their products, ensuring that environmental and social costs are internalised rather than externalised to communities like those in Bangladesh.[53] Paired with ESPR, which sets requirements for durability, repairability and sustainability at the design stage, these tools are intended to reduce waste, extend product lifespans

[48] D'Alisa, G, Federico Demaria, and Giorgos Kallis. 2015. *Degrowth: A Vocabulary for a New Era*. Routledge.

[49] Swedwatch. 2024. Paying the price for fashion: securing a living wage for Bangladesh's garment workers. Available at: https://swedwatch.org/wp-content/uploads/2024/11/briefsecuring-living-wagesnov-201124.pdf (Accessed: 20 August 2025).

[50] Hickel, J. 2021. Less Is More: How Degrowth Will Save the World. London: William Heinemann; Bhambra, G. K. 2022. A Decolonial Project for Europe. *JCMS: Journal of Common Market Studies, 60*(2): 229–244. https://doi.org/10.1111/jcms.13310

[51] Suraci, O. 2021. The Best-Dressed Polluter – Regulation and Sustainability in the Fashion Industry. *Hastings Environmental Law Journal, 27*(2): 225.

[52] European Commission. 2022. First Circular Economy Package including Regulation proposal on Sustainable Products and Sustainable Textiles Strategy. Available at: https://ec.europa.eu/newsroom/env/items/742027/en (Accessed: 20 August 2025); Brown, A., and P. Börkey. 2024. Extended producer responsibility in the garments sector. *OECD Environment Working Papers, 253*. https://doi.org/10.1787/8ee5adb2-en

[53] Suraci, O. 2021. The Best-Dressed Polluter – Regulation and Sustainability in the Fashion Industry. *Hastings Environmental Law Journal, 27*(2): 225.

and drive circularity.[54] In practice, however, EPR often becomes narrowly technocratic, focused on downstream waste collection and recycling,[55] while Producer Responsibility Organisations, which are the industry bodies that manage EPR schemes, tend to prioritise logistical efficiency.[56] Grassroots and labour actors are frequently excluded from these processes, and the crucial link between eco-design, consumption, reduction and fair working conditions in the Global South is sidelined.[57] Consequently, the promises of EPR to reshape production and consumption are diluted into an end-of-pipe solution that leaves overproduction and overconsumption intact.

For Bangladesh, this means that circularity targets in the Global North, while progressive on paper, risk marginalising local recycling economies or shifting environmental burdens without addressing trade inequalities and structural dependency.[58] Without binding commitments to reduce the total volume of production and foster fair trade, ESPR and EPR risk reinforcing the 'green growth' paradigm that prioritises European material security, for which there is no empirical evidence, over a just, multilateral transition.[59] Beyond economic displacement, they can concentrate environmental harms locally (including air and soil contamination, toxic residues from informal recycling and water pollution from effluents and runoff), thereby compounding public-health and ecological impacts.[60] Furthermore, evidence from Bangladesh shows how these top-down models can exacerbate

[54] Mallick, P. K, *et al.* 2024. Designing and operationalising extended producer responsibility under the EU Green Deal. *Environmental Challenges, 16.* https://doi.org/10.1016/j.envc.2024.100977

[55] Lifset, R. *et al.* 2023. Restoring the incentives for eco-design in extended producer responsibility: The challenges for eco-modulation. *Waste Management,* 168: 189–201. https://doi.org/10.1016/j.wasman.2023.05.033

[56] Mayanti, B. and Helo, P. 2023. Circular economy through waste reverse logistics under extended producer responsibility in Finland. *Waste Management & Research, 42*(1): 59–73. https://doi.org/10.1177/0734242X231168801

[57] Suraci, O. 2021. The Best-Dressed Polluter – Regulation and Sustainability in the Fashion Industry. *Hastings Environmental Law Journal, 27*(2): 225.

[58] Swedwatch. 2024. Paying the price for fashion: securing a living wage for Bangladesh's garment workers. Available at: https://swedwatch.org/wp-content/uploads/2024/11/briefsecuring-living-wagesnov-201124.pdf (Accessed: 20 August 2025).

[59] Hickel, J. and Kallis, G. 2020. Is Green Growth Possible? *New Political Economy, 25*(4): 469–486. https://doi.org/10.1080/13563467.2019.1598964

[60] Coglianese, C. and Nash, J. 2001. *Regulating from the Inside: Can Environmental Management Systems Achieve Policy Goals?* RFF Press.

existing burdens.[61] When Western brands or governments impose new environmental requirements without supporting local capacities, the costs fall on suppliers who already operate with razor-thin margins.[62] Factories may be forced to absorb new costs or risk losing contracts, passing pressures down to workers through unpaid overtime, job loss, or intensified production demands.[63] In effect, such policies shift responsibility without redistributing power.

This dynamic exposes a key flaw in supposedly 'holistic' approaches: if they do not meaningfully include producer countries in governance and design, they risk perpetuating colonial trade logics.[64] Sustainability becomes a one-way flow of obligations and standards, rather than a collaborative process grounded in local knowledge, needs and priorities. This also echoes the failures and limitations of the reactive, sectoral policy cycle discussed in section 3. If change remains contingent on Northern interests, underlying global inequalities remain intact.[65] A genuine just transition framework must therefore do more than integrate environmental and social goals on paper. It must ensure that policy design is participatory and multilateral, giving voice and agency to producer countries, instead of them being passive recipients.[66] This means moving beyond mere consultation to co-creation and allowing local unions, civil society

[61] Alam, Hasnat, Patrick Schröder, and Masrur Reaz. 2024. International Circular Economy Policy Trends - Implications for Bangladesh's Circular Garment and Textile Policy Development. https://www.switchtocircular.eu/sites/default/files/publications/International%20circular%20economy%20policy%20trends_implications%20for%20Bangladesh_pub_web.pdf Last Accessed 20 August 2025.

[62] Alam, H. *et al.* 2024. International Circular Economy Policy Trends - Implications for Bangladesh's Circular Garment and Textile Policy Development. Available at: https://www.switchtocircular.eu/sites/default/files/publications/International%20circular%20economy%20policy%20trends_implications%20for%20Bangladesh_pub_web.pdf

[63] ILO. 2023. The Rana Plaza disaster ten years on: What has changed? Available at: https://webapps.ilo.org/infostories/en-GB/Stories/Country-Focus/rana-plaza#intro (Accessed: 20 August 2025).

[64] Bhambra, G. K. 2022. A Decolonial Project for Europe. *JCMS: Journal of Common Market Studies, 60*(2): 229–244. https://doi.org/10.1111/jcms.13310

[65] Choudhury, M. and Haque, C.M. 2024. Disaster management policy changes in Bangladesh: Drivers and factors of a shift from reactive to proactive approach. *Environmental Policy and Governance, 34*(5): 445–462. https://doi.org/10.1002/eet.2094

[66] Hickel, Jason. 2021. *Less Is More: How Degrowth Will Save the World.* London: William Heinemann.

and small producers to shape what sustainability means contextually.[67] Framing this through an energy-justice perspective is useful because it foregrounds three (or four) mutually reinforcing concerns: (i) the fair distribution of benefits and burdens (distributive justice), (ii) the right to meaningful participation in decision-making (procedural justice) and (iii) the recognition of historically marginalised groups (recognition justice & restorative justice) and translates them into concrete safeguards (e.g. mandated co-design processes, equitable benefit-sharing mechanisms, transparency and accountability requirements and health-based environmental standards). Doing so makes it far more likely that transitions will reduce, rather than concentrate, local environmental and public-health harms.[68]

Some emerging governance models provide inspiration. Polycentric arrangements, trialled in parts of Vietnam's textile sector, bring together local governments, workers, brands and community groups to jointly develop and monitor standards.[69] This shared responsibility can help balance the power asymmetries inherent in current global supply chains.[70] It also aligns with the just transition principle that workers and communities must not bear the costs of systemic change alone.[71] Another critical lesson is that holistic approaches must confront the sheer scale of production and consumption. Technical improvements such as eco-design and recycling cannot replace binding commitments to reduce material throughout, especially in the Global North.[72] Without this, efficiency gains are easily offset by the 'rebound effect.'[73] This risks locking countries like Bangladesh into the same exploitative position of supplying cheap garments for unsustainable consumption elsewhere. Crucially, this is not only a matter of global justice but of immediate local harm: textile production

[67] Bhambra, G. K. 2022. A Decolonial Project for Europe. *JCMS: Journal of Common Market Studies, 60*(2): 229–244. https://doi.org/10.1111/jcms.13310

[68] Sovacool, B. K., and Dworkin, M.H. 2015. Energy justice: Conceptual insights and practical applications. *Applied Energy, 142*: 435–444. https://doi.org/10.1016/j.apenergy.2015.01.002

[69] Suraci, O. 2021. The Best-Dressed Polluter – Regulation and Sustainability in the Fashion Industry. *Hastings Environmental Law Journal, 27*(2): 225.

[70] Bhambra, G. K. 2022. A Decolonial Project for Europe. *JCMS: Journal of Common Market Studies, 60*(2): 229–244. https://doi.org/10.1111/jcms.13310

[71] Hickel, J. 2021. *Less Is More: How Degrowth Will Save the World*. London: William Heinemann.

[72] D'Alisa, G. *et al.* 2015. *Degrowth: A Vocabulary for a New Era*. Routledge.

[73] Polimeni, J. M. *et al.* 2008. *The Jevons Paradox and the Myth of Resource Efficiency Improvements*. London: Earthscan.

in exporting regions frequently produces large volumes of wastewater, releases untreated chemical effluents, contributes to soil and groundwater contamination and generates hazardous solid wastes, all of which impose health, livelihood and ecosystem costs on local communities.[74]

In short, holistic policy frameworks are only as transformative as the assumptions about the political economy they operate with. Without confronting the root drivers, they risk becoming new tools for technocratic or neocolonial control. In contrast, this chapter argues, a just transition insists that sustainability must be built on solidarity, redistribution and participation. For the global fast fashion industry, this means grounding circular economy or extraterritorial due diligence initiatives that may affect producer countries like Bangladesh[75] in binding commitments to reduce production and consumption, guarantee decent livelihoods and transfer resources to support structural changes in these countries. Only then can holistic approaches move beyond rhetoric and address the systemic injustices that crises like the Rana Plaza collapse laid bare.

5.5 Degrowth Within a Just Transition Framework

While holistic governance frameworks aim to address the interconnected social and environmental crises in fast fashion, they often stop short of the foundational economic drivers. As Sections 1–4 have illustrated, piecemeal improvements have helped mitigate some harms, yet they remain embedded in a global system dependent on overproduction and overconsumption.[76] Degrowth directly confronts this dependence, especially as the foundation of a part of a just transition framework. In one sentence, degrowth can be defined as '[…] a planned reduction of energy and resource use designed to bring the economy back into balance with the living world in a way that reduces inequality and improves human

[74] Coglianese, C. and Nash, J. 2001. *Regulating from the Inside: Can Environmental Management Systems Achieve Policy Goals?* RFF Press.

[75] Afrose, M. 2024. EU Due-Diligence Directive for Corporate Sustainability: Consequences for Bangladesh. Bangladesh Foreign Trade Institute. Available at: https://www.bfti.org.bd/blog-details/11 (Accessed: 20 August 2025).

[76] D'Alisa, G. *et al.* 2015. *Degrowth: A Vocabulary for a New Era.* Routledge; Fletcher, K. and Tham, M. 2019. Earth Logic: Fashion Action Research Plan. Available at: https://katefletcher.com/wp-content/uploads/2019/10/Earth-Logic-plan-FINAL.pdf (Accessed: 20 August 2025).

well-being.'[77] It offers a radical alternative to the growth-dependent model that relies on cheap labour, lax environmental standards and the exploitation of countries like Bangladesh to satisfy consumption in the Global North.[78] Rather than treating sustainability as an efficiency problem in need of technical solutions, degrowth calls for a managed, intentional and equitable downscaling of material and energy throughput, alongside programmes that protect livelihoods and redistribute resources.[79]

For the fast fashion industry, the case for degrowth is clear. Despite efficiency gains, total production volumes continue to rise, undermining the environmental improvements of sustainable design or recycling.[80] The 'rebound effect' perpetuates demand for cheap garments, keeping Bangladesh's economy locked into precarious, low-margin supply chains.[81] To have a just transition, this structural dynamic cannot be ignored. Practically, degrowth-oriented policies would aim to reduce unnecessary and environmentally harmful consumption in the Global North while ensuring social protections for workers in producer countries, especially during the period of industrial transformation. Measures could include legally binding caps on resource extraction for textiles, bans on planned obsolescence and taxes on overproduction.[82] These would need to be paired with meaningful investments into repair and reuse infrastructures, supporting alternative forms of work that are not reliant on exploitative supply chains.[83]

[77] Hickel, J. 2021. What does degrowth mean? A few points of clarification. *Globalizations, 18*(7): 1105–1111. https://doi.org/10.1080/14747731.2020.1812222

[78] Hickel, J. and Kallis, G. 2020. Is Green Growth Possible? *New Political Economy, 25*(4): 469–486. https://doi.org/10.1080/13563467.2019.1598964

[79] Latouche, S. 2009. Farewell to growth. Cambridge: Polity Press; Vanderheiden, Steve. 2020. *Environmental Political Theory*. Cambridge: Polity Press.

[80] Hertwich, E. G. 2005. Consumption and the Rebound Effect: An Industrial Ecology Perspective. *Journal of Industrial Ecology, 9*(1–2): 85–98. https://doi.org/10.1162/1088198054084635; Polimeni, J. M. *et al.* 2008. *The Jevons Paradox and the Myth of Resource Efficiency Improvements*. London: Earthscan.

[81] Swedwatch. 2024. Paying the price for fashion: securing a living wage for Bangladesh's garment workers. Available at: https://swedwatch.org/wp-content/uploads/2024/11/briefsecuring-living-wagesnov-201124.pdf (Accessed: 20 August 2025).

[82] Jackson, T. 2016. *Prosperity Without Growth: Foundations for the Economy of Tomorrow*. Oxford: Taylor & Francis Group.

[83] Fletcher, K. 2010. Slow Fashion: An Invitation for Systems Change. *Fashion Practice, 2*(2): 259–265. https://doi.org/10.2752/175693810X12774625387594

Critically, degrowth does not take away the Global South's right to develop. On the contrary, it recognises that countries such as Bangladesh have been structurally positioned as low-cost producers for the Global North's unsustainable consumption patterns.[84] A degrowth framework highlights the importance of challenging the dominant liberal political philosophy that equates economic growth with progress.[85] Yet, the degrowth literature is explicit in targeting wealthy, overdeveloped economies whose ecological footprints vastly exceed planetary boundaries. Far from denying countries in the Global South the right to develop and improve livelihoods, degrowth seeks to create the space for ecological as well as economic sovereignty by reducing extractive pressures.[86] It is not about denying the right to flourish but about disentangling this progress from dependency on extractive demand. This means recognising the 'imperial mode of living' that drives exploitative labour conditions, environmental degradation and resource depletion in producer countries while extracting the profit to the Global North.[87]

A degrowth-orientated just transition would also be able to address the inherent power imbalance in global supply chains. Currently, Bangladeshi suppliers have little bargaining power over large international buyers who dictate prices and deadlines.[88] This imbalance contributes to dangerous working conditions and environmental damage, as demonstrated by the Rana Plaza collapse and other disasters.[89] By targeting the consumption side, degrowth shifts the responsibility back onto the Global North, where

[84] Bhambra, G. K. 2022. A Decolonial Project for Europe. *JCMS: Journal of Common Market Studies, 60*(2): 229–244. https://doi.org/10.1111/jcms.13310

[85] Vanderheiden, S. 2020. *Environmental Political Theory*. Cambridge: Polity Press.

[86] Demaria,F. *et al.* 2013. What is Degrowth? From an Activist Slogan to a Social Movement. *Environmental Values, 22*(2): 191–215. https://doi.org/10.2307/23460978

[87] Smith, J. 2016. *Imperialism in the Twenty-First Century: Globalization, Super-Exploitation, and Capitalism's Final Crisis*. New York: Monthly Review Press; Brand, U. and Wissen, M. 2021. *The Imperial Mode of Living: Everyday Life and the Ecological Crisis of Capitalism*. London: Verso.

[88] Alam, H. *et al.* 2024. International Circular Economy Policy Trends - Implications for Bangladesh's Circular Garment and Textile Policy Development. Available at: https://www.switchtocircular.eu/sites/default/files/publications/International%20circular%20economy%20policy%20trends_implications%20for%20Bangladesh_pub_web.pdf

[89] Fletcher, K. and Tham, M. 2019. Earth Logic: Fashion Action Research Plan. Available at: https://katefletcher.com/wp-content/uploads/2019/10/Earth-Logic-plan-FINAL.pdf (Accessed: 20 August 2025).

policy levers such as consumption taxes and advertising restrictions can curb the demand for disposable garments.

Moreover, degrowth can provide the normative foundation for aligning existing tools, like due diligence laws or circular economy strategies, with systemic transformation. Without such a normative foundation, these tools risk becoming technocratic fixes that fail to address the scale of the problem.[90] For example, eco-design standards can help extend product lifespans, but unless total production is reduced, overall environmental impacts continue to rise. Leaning on degrowth ensures that sustainability is about sufficiency, not just efficiency, because 'unless brands commit to reducing the number of new items they produce, they are only expanding the 'circle' [...].'[91]

Opponents of degrowth argue that downscaling production would threaten jobs and livelihoods in export-orientated economies like Bangladesh. The jobs are already deeply insecure and dependent on conditions that produce regular human rights and environmental violations.[92] The just transition must therefore come with a guarantee of income protection, upskilling and pathways to alternative livelihoods. This would involve strengthening social safety nets, promoting local industries beyond textiles and supporting democratic worker cooperatives.[93] This reorientation is essential in overcoming the limitations of reactive, sectoral policymaking. As Section 3 illustrates, policy reforms often follow high-profile crises, while only addressing immediate, tangible issues like building safety, leaving aside deeper drivers of exploitation.[94] An approach inspired and built on the principles of degrowth understands these risks and anticipates

[90] Mallick, P.K. 2024. Designing and operationalising extended producer responsibility under the EU Green Deal. *Environmental Challenges, 16*: 100977. https://doi.org/10.1016/j.envc.2024.100977

[91] Fashion Revolution. 2023. Fashion Transparency Index. Available at: https://www.fashionrevolution.org/fashion-transparency-index-2023/ (Accessed: 20 August 2025).

[92] Swedwatch. 2024. Paying the price for fashion: securing a living wage for Bangladesh's garment workers. Available at: https://swedwatch.org/wp-content/uploads/2024/11/briefsecuring-living-wagesnov-201124.pdf (Accessed: 20 August 2025).

[93] Jackson, T. 2016. *Prosperity Without Growth: Foundations for the Economy of Tomorrow*. Oxford: Taylor & Francis Group.

[94] Choudhury, M. and Haque, C.E. 2024. Disaster management policy changes in Bangladesh: Drivers and factors of a shift from reactive to proactive approach. *Environmental Policy and Governance, 34*(5): 445–462. https://doi.org/10.1002/eet.2094

future issues by tackling the underlying political economy that incentivises unsafe conditions and destructive production models.[95]

Degrowth also insists that sustainability governance must be multilateral and inclusive. Unilateral frameworks, such as EU due diligence directives, risk reproducing colonial asymmetries since they fail to meaningfully include the voices from producer countries. A just degrowth transition would embed participatory and recognitional mechanisms with the intention of giving agency to affected workers and communities to shape the policies that impact them directly. Polycentric governance models, which have already been trialled in parts of Asia, demonstrate the potential of local ownership when diverse stakeholders share power and accountability.[96] Practically, this means that any strategy to downscale fast fashion must come with robust support for countries like Bangladesh. Debt relief, climate finance and trade reforms are all critical to giving producer countries the fiscal and institutional capacity to transition without deepening poverty or inequality.[97] In the garment sector, this could involve shifting from mass export production to higher-value, smaller-scale enterprises focused on quality, sustainability and local markets.

Importantly, however, degrowth is not a silver bullet that will eradicate all our contemporary problems. It is not meant to replace regulatory tools like due diligence laws, LCA, or corporate accountability frameworks, all of which have had undeniably positive impacts. Instead, it offers a normative foundation and orientation that can align and strengthen them. Without a guiding commitment to sufficiency, sustainability and justice, even well-intentioned policies risk falling into the traps of green capitalist logics, where production grows, branding improves and underlying issues and injustices continue to exist.[98] Degrowth can provide the political and normative framing that is needed to move beyond green consumerism and towards the structural change mentioned throughout this chapter. Degrowth can help situate the existing and future interventions within

[95] Hickel, J and Kallis, G. 2020. Is Green Growth Possible? *New Political Economy, 25*(4): 469–486. https://doi.org/10.1080/13563467.2019.1598964

[96] Fasting, S. *et al.* 2021. Climate Governance and Agriculture in Southeast Asia: Learning From a Polycentric Approach. *Frontiers in Political Science, 3.* https://doi.org/10.3389/fpos.2021.698431

[97] Hickel, J. 2021. *Less Is More: How Degrowth Will Save the World.* London: William Heinemann.

[98] Corvellec, H. Stowell, Alison, F. and Johansson, N. 2022. Critiques of the circular economy. *Journal of Industrial Ecology, 26*(2): 421–432. https://doi.org/10.1111/jiec.13187

a broader political-economic transformation. This transformation would be one that questions not only how garments are made, but also why, for whom and how many. It insists that ecological and social sustainability cannot be pursued without addressing global inequality and Northern overconsumption.

In short, degrowth is not a niche radicalism. It is increasingly recognised as a necessary framework for simultaneously addressing overlapping issues of climate breakdown, biodiversity loss and social injustice.[99] For the fashion industry, it can offer a path out of extractivist logic and towards a system built on care, creativity and sufficiency. A degrowth transition will not be easy, perfect, or quick but without it, the currently pursued reforms may remain insufficient to tackle contemporary problems.

5.6 Conclusion

This chapter has illustrated that the global fast fashion industry's reliance on the exploitation of labour and the environment demands more than reactive, sectoral policy responses if a just transition can indeed become a reality. The case of Bangladesh, epitomised by the Rana Plaza collapse, shows that crisis-driven reforms often improve immediate working conditions but leave out any concerns about the underlying extractive dynamics. Holistic governance frameworks and programmes, such as LCA, due diligence laws and circular economy strategies, represent a progressive step in the right direction by recognising the interdependence of social, environmental and economic systems. Yet, as this chapter has argued, these approaches risk perpetuating colonial power asymmetries when implemented unilaterally and fail to consider the aspects of overconsumption and overproduction.

A just transition framework built on the principles of degrowth can help reconcile these issues. It confronts the systemic drivers of unsustainability at their political-economic roots, calling for deliberate reductions in material throughput and demand in the Global North while simultaneously securing economic and social protections for workers in the producer

[99] Hickel, J. and Kallis, G. 2020. Is Green Growth Possible? *New Political Economy*, *25*(4): 469–486. https://doi.org/10.1080/13563467.2019.1598964; IPCC. 2022. Climate Change 2022: Mitigation of Climate Change. Available at: https://www.ipcc.ch/report/ar6/wg3/downloads/report/IPCC_AR6_WGIII_FullReport.pdf (Accessed: 20 August 2025).

countries like Bangladesh. This means moving beyond technocratic fixes and regulatory checklists to building multilateral, participatory governance structures that give agency and recognition to affected communities and ensuring that efforts towards more sustainable ways of living do not reproduce neocolonial power asymmetries.

The insights from Bangladesh's experience underline that a truly just transition must be anticipatory, transformative and globally coordinated. It must address the demand side of fast fashion alongside production practices and redistribute the costs and benefits of change equitably. For policymakers, this requires bringing together environmental targets with social protections and global trade reforms. For brands, it means taking responsibility not only for how clothes are made but also for how many are made and why. For communities, it calls for political organising and demanding inclusion in shaping what a just and sustainable future for fashion should look like.

In sum, a just transition framework that integrates holistic governance with the principles of degrowth offers a credible path towards overcoming the failures and limitations of reactive, sectoral policymaking. Only by addressing the root drivers of social and environmental injustice can the fashion industry truly transform, in Bangladesh and beyond.

CHAPTER 6

Towards 'Just' Mining: From Restoration to Restorative Justice

Francisca Soto Monteverde, Clelia Zardini, Tareq Yacoub Helou, Carlotta Terhorst, and Adetokunbo Alase

Abstract The global decarbonisation process will depend significantly on minerals and metals, as the transition away from fossil fuel technologies requires substantially greater quantities of these resources. Historically, Latin America (LATAM) has served as a major site for mineral extraction, a process that has contributed to considerable economic growth. However, this development has come at a high environmental cost, leading to severe

F. Soto Monteverde
Centre for Climate Law and Governance, Dickson Poon School of Law, King's College London, London, United Kingdom
e-mail: francisca.soto_monteverde@kcl.ac.uk

C. Zardini
Università Statale di Milan, Milan, Italy

T. Yacoub Helou
University of São Paulo and King's College London, London, United Kingdom

C. Terhorst
Socio-Ecological Transformation (SET) research group of the Linz Institute for Transformative Change, Linz, Austria

A. Alase
Faculty of Law, University of Manitoba, Winnipeg, Canada

A. Phillips, L. Kaschny (eds.), *Global Perspectives on the Circular Economy in the Just Transition*, Just Transitions,
https://doi.org/10.1007/978-3-032-18342-2_6

ecological degradation and serious public health issues, which has been understood as 'extractivism' by relevant academic literature. In this context, LATAM faces a critical dilemma: how can the region maximise the economic benefits of mining while simultaneously safeguarding the environment and upholding human rights according to fairness principles? We propose the application of a restorative justice emphasis to the mining sector in LATAM as a framework capable of addressing historical injustices and fostering equitable outcomes for affected communities. Building on the 2024 judgement of the Inter-American Court of Human Rights (IACHR) in the case of Inhabitants of La Oroya v. Peru, this paper analyses how a restorative justice approach can contribute to the remediation and prevention of environmental harm and public health issues arising from extractive activities in the region. The relevance of this case lies in its contribution to the advancement of environmental justice within the extractive industry, as it constitutes the first contentious case to ground its legal reasoning in the right to live in a healthy environment. It thereby demonstrates how a restorative justice approach can help to prevent serious environmental harm and associated human rights violations in the context of mining for the energy transition.

Keywords Extractivism; Minerals • Latin America • Just Transition • Restorative Justice • Human Rights • Healthy Environment • Inter-American Human Rights System

6.1 Introduction

The mining sector in Latin America is often associated with socio-environmental conflict.[1] Especially in the context of a capitalistic polycrisis (i.e. '*world historical condition of global turbulence driven by the intersecting crises of capitalism, energy, and the earth system*'),[2] extractivist practices are heavily criticised.[3] Despite this, the extraction of raw minerals such as

[1] Arboleda, M. 2020. Planetary Mine: Territories of Extraction under Late Capitalism. Verso; Sabatini, C. and Binetti, B. 2024. State Capacity, Mining and Community Relations in Peru: How Sustained Multi-Sectoral Dialogue Can Help Secure the Country's Role in Global Supply Chains. Chatam House. Available at: https://www.chathamhouse.org/sites/default/files/2024-12/2024-12-02-state-capacity-mining-community-relations-peru-sabatini-binetti.pdf (Accessed: 20 October 2025)

[2] Albert, M. J. 2024. *Navigating the Polycrisis: Mapping the Futures of Capitalism and the Earth*. The MIT Press.

[3] Veltmeyer, H. 2022. *Extractivism and Beyond*. The Extractive Industries and Society. https://doi.org/10.1016/j.exis.2022.101132

copper or lithium remains essential for combating climate change globally. This is because low-carbon technologies require significantly larger quantities of these resources compared to fossil-fuel technologies.[4] Countries gifted with these resources have already experienced economic gains from rising demand, yet the potential for a sustainable extraction in accordance with principles of justice remains largely and historically unrealised in LATAM.

Under such scenario, LATAM is facing a crucial dilemma: how can the region enhance the economic benefits of mining while, at the same time, protecting the environment and upholding human rights? A pivotal moment in addressing these issues[5] occurred in March 2024 when the IACHR delivered a landmark decision in the case of *La Oroya v. Peru*,[6] where the state of Peru was declared responsible for violating the right to a healthy environment, health and life of La Oroya inhabitants and children's rights, caused by the toxic activities of the mining smelter complex deployed by the multinational corporation Doe Run. Peru was found responsible for failing to regulate and monitor the activities of companies operating in the La Oroya metallurgical complex – activities it knew to be toxic and harmful to public health. Particularly relevant is the Court's application of the non-regression principle, as the Peruvian state had issued a supreme decree that weakened air quality standards. This offers an important lesson for future administrations in presidential regimes, which are the most prevalent in LATAM.

While this decision represents progress in terms of environmental justice in the region, it also raises questions about its broader implications for the mining sector in the context of the global energy transition. This chapter addresses some of them, starting from investigating whether the Court's decision can prevent similar occurrences in other countries or areas of Peru. Moreover, it discusses the responsibility of international

[4] Morton Turner, J. 2022. Charged: A History of Batteries and Lessons for a Clean Energy Future University of Washington Press.

[5] At the time of writing this article, the Advisory Opinion on Climate Emergency and Human Rights had not yet been issued. See State obligations on Climate Emergency and Human Rights (Advisory Opinion) Inter-American Court of Human Rights (29 May 2025) AO-32/25 Series A No 32.

[6] Case of La Oroya Population v. Peru. (Judgment of 27 November 2023, on Preliminary Objections, Merits, Reparations and Costs) Inter-American Court of Human Rights Ser C No 511.

corporations profiting from extractive practices and its role in challenging the prevailing unsustainable development strategies in Peru and across Latin America.

To address these challenges, this article proposes the application of the restorative justice principle to the mining sector in LATAM, which remains vital to the global energy transition. Heffron and Hazrati suggest that integrating restorative justice into energy transition can redress historical injustices while fostering equitable outcomes for affected communities.[7] By prioritising reconciliation and community participation, this approach offers a framework that not only mitigates environmental impacts but also advances social justice and contributes to sustainable development. This chapter builds on political economy and critical legal literature, aiming at making a conceptual contribution grounded in legal case analysis. The goal of the chapter is to contribute to an effective and critical approach to the 'just transition' and to lay the groundwork for a new legal framework and development policy for the mining industry in LATAM.

6.2 Advancing a Restorative Justice Approach in the Mining Sector

The theory of restorative justice in the mining industry for the energy transition implies the restoration of the environment and communities affected by mining projects to the state and quality that they had before the commencement of these activities, to their original state.[8] More fundamentally, restorative justice emphasises that these considerations should be taken into account during the entire mineral value chain or life cycle of the mineral products,[9] for instance, lithium-ion batteries.[10] Within the just transition scholarship, restorative justice encapsulates a broader scope,

[7] Heffron, R.J. and Hazrati, M. 2024. Conceptualising Restorative Justice in the Energy Transition. In: Heffron, R.J. and Hazrati,M. (Eds.) The Rise of Restorative Justice in the Energy Transition and for Climate Mitigation. Springer Nature Switzerland.

[8] Nakanwagi, S. 2024. Critical Minerals, Sustainability, and the Energy Transition in the Global South: A Justice Perspective. Hart. 135–136.

[9] Ibid.

[10] Zengin, O.Z. 2024. The Energy Justice Imperative for Clean Energy Storage Alternatives In: Heffron, R.J. and de Fontenelle, L. (Eds). *The Power of Energy Justice & the Social Contract.* Just Transitions. Palgrave Macmillan, Cham. 222.

including the restoration of people, nature and the planet to their original state before the harm caused by extractive activities.[11]

While restorative justice originates in the field of criminal justice – seeking to provide remedies and reparations to victims beyond the sole focus on individual punishment – [12] in the context of environmental problems,[13] it aims both to repair damage and to prevent further environmental degradation alongside the reparation of broken relationships in sites of socio-environmental conflict.[14]

Likewise, the origins of restorative justice emphasise a broader understanding of victims of wrongdoing, encompassing the society as a whole.[15] Thus, communicative actions that acknowledge, recognise and repair the harm are essential. These could include dialogue, truth-telling, inclusivity, apologies and material payments. Restorative justice also highlights the importance of responsibility in addressing harm and promoting healing. It asserts that those who have caused harm should take accountability for their actions by recognising and acknowledging the impact of their behaviour. With all, in situations of environmental damages caused by mining extraction, it is necessary to recognise the power imbalances between corporations and the people, or the people and state parties.[16] Hence, strategies such as an impartial body might be needed to mediate between parties.

The potential of restorative justice approaches lies in their utility for the prevention, management, and remediation of environmental harm. These aims shall be deployed across all stages of extraction, involving environmental impact assessment, project operation, project closure and

[11] Hazrati, M. and Heffron, R.J. 2021. Conceptualising Restorative Justice in the Energy Transition: Changing the Perspectives of Fossil Fuels. *Energy Research & Social Science, 78.* https://doi.org/10.1016/j.erss.2021.102115

[12] Ibid.

[13] Fisher, E. *et al.* 2019. What Is Environmental Law? In: Fisher, E., Lange, B. and Scotford, E. (Eds.) *Environmental Law*. Oxford University Press. *19.*

[14] Forsyth, M. *et al.* 2022. Environmental Restorative Justice: An Introduction and an Invitation. In: Pali, B., Forsyth, M. and Tepper F. (eds). *The Palgrave Handbook of Environmental Restorative Justice.* Springer International Publishing. *3.*

[15] Menkel-Meadow, C. 2007. Restorative Justice: What Is It and Does It Work? *Georgetown Law Faculty Publications and Other Works*, *2.*

[16] Forsyth, M. *et al.* 2022. Environmental Restorative Justice: An Introduction and an Invitation. In: Pali, B., Forsyth, M. and Tepper F. (eds). *The Palgrave Handbook of Environmental Restorative Justice.* Springer International Publishing. *3.*

end-of-life of the mining value chain. This could include early actions with the existing infrastructure, such as seaports, road networks, traffic pressures and other businesses that use the infrastructure that mining projects feed off. As the project goes into operation, it is necessary to continue engaging with local stakeholders and consider changes that highlight newly recognised hazards and risks. An example would be to analyse emerging data that identifies new forms of environmental harm and respond to this new information in a way that listens to the voices of community and business stakeholders as the project continues. As can be seen, these procedures and actions can also be informed by an energy justice lifecycle framework, which treats the entire system of energy production, transportation, and consumption as a single analytical category, thereby enabling holistic approaches to energy justice.[17]

As presented, restorative justice is a compelling approach to explore and implement in mining projects for avoiding extractivism. First, because it includes that the entire lifecycle of mining is open, participatory and inclusive of a network of interdependent stakeholders who would or might be affected by energy transition projects. This should not only be demonstrated in explicit mention of procedural decision-making but also be score to the entire process of extraction from project assessment to closure. Specifically, the four stages in the value chain are identified as the investment and production stage, processing and refining stage, consumption stage, and waste management and mine closure.[18] Second, because it pursues specific outcomes, namely the prevention and remediation of harm to the environment, communities, and society.[19] In this way, it can help avert the severe environmental degradation commonly associated with mining projects in LATAM.

[17] Heffron, R.J. and McCauley, D. 2017. The Concept of Energy Justice across the Disciplines. *Energy Policy*, *105*: 658–667. https://doi.org/10.1016/j.enpol.2017.03.018; Mejía-Montero, A. *et al.* 2021. Grounding the Energy Justice Lifecycle Framework: An Exploration of Utility-Scale Wind Power in Oaxaca, Mexico. *Energy Research & Social Science*, *75*. https://doi.org/10.1016/j.erss.2021.102017

[18] Nakanwagi, S. 2024. *Critical Minerals, Sustainability, and the Energy Transition in the Global South: A Justice Perspective*. Hart Publishing. 12.

[19] Heffron, R.J. and McCauley, D. 2017. The Concept of Energy Justice across the Disciplines. *Energy Policy*, *105*: 658–667. https://doi.org/10.1016/j.enpol.2017.03.018

6.3 Extractivism in Latin America: A Critical Review of Its Economic, Social and Environmental Impacts

Extractivism constitutes a fundamental activity within capitalist economies, wherein natural goods, sourced from territories, mountains, and mines, are extracted and transformed into marketable goods that generate surplus value through technological processes resulting in the production of commodities.[20] Although the term *extractivism* appeared in scholarly literature during the 1970s and was formalised in the 1980s,[21] it has since evolved into a critical lens through which to examine capitalist development models reliant on transnational corporations, multilateral financial institutions, and complicit state actors. Contemporary characterisations of extractivism further highlight that most of the wealth is produced at the end of the value chain, thereby marginalising workers and local populations positioned at the beginning of the extractive process from both the distribution of wealth and access to economic power.[22]

6.3.1 Latin American Extractivism Literature

In LATAM, extractivism is further characterised as a long-term process involving multinational corporations that are allowed to extract, appropriate, and export natural resources, supported by national policies that promote and facilitate this model of development.[23] Hence, the extraction of natural resources has long served as a central development strategy – pursued by both governments and corporations – as a means of expanding productive capacity and enabling capital accumulation. As Fernandes observes, in regions where extraction occurs, both the workforce and the

[20] Chagnon, C. *et al.* 2022. From Extractivism to Global Extractivism: The Evolution of an Organizing Concept. *The Journal of Peasant Studies, 49*(4): 760–792. https://doi.org/10.1080/03066150.2022.2069015

[21] Durante, F. *et al.* 2021. Extraction and Extractivisms: Definitions and Concepts. In: Shapiro, J. and McNeish, J.A. (Eds) *Our Extractive Age: Expressions of Violence and Resistance.* Routledge.

[22] Veltmeyer, H. 2022. Extractivism and Beyond: Latin America debates. *The Extractive Industries and Society, 11.* https://doi.org/10.1016/j.exis.2022.101132

[23] Arias-Loyola, M. *et al.* 2022. Beyond the Resource Curse: The Redistributive Challenge of Sustainable Resource-Led Development in Australia, Chile and Zambia. *The Extractive Industries and Society, 11.* https://doi.org/10.1016/j.exis.2022.101084

localities themselves are reshaped, with towns and communities transformed into so-called 'sacrifice zones.'[24] This term refers to geographical areas which disproportionately absorb the environmental burdens of industrial and extractive activities, while the benefits of such activities are realised elsewhere and by different populations.[25] The unequal allocation of environmental harms and economic gains inherent in the creation of such zones is a core concern of distributive justice, dealing with the allocation of costs and rewards of development.

With the rise of the neoliberal world order and the commodities boom of the 1990s, extractivism became the subject of extensive debate, increasingly criticised for contributing to environmental degradation and exacerbating social and economic inequalities.[26] This scenario became more complex with the emergence of the 'pink tide' in LATAM, during which left-leaning governments paradoxically embraced extractivism as a development strategy, channelling revenues into neo-developmentalist policies.[27] In the early 2000s, a 'left turn' in LATAM initially was informed by alternative approaches to environmental governance – including the constitutional recognition of the 'Rights of Nature' in Ecuador – yet these shifts were not accompanied by efforts to decouple national economies from neoliberal models of extractivism.[28] For explaining these contradictions, Svampa introduced the concept of neo-extractivism, indicating a development paradigm in LATAM marked by large-scale and intensified exploitation of natural goods for its trading and export, alongside

[24] Fernandes, S. 2024. Just Means 'Just' Everywhere: How Extractivism Stands in the Way of an Internationalist Paradigm for Just Transitions. *International Journal of Politics, Culture, and Society*, *37*(4): 493–511. https://doi.org/10.1007/s10767-024-09475-4

[25] Juskus, R. 2023. Sacrifice Zones: A Genealogy and analysis of an environmental justce concept. *Environmental Humanities*, *15*(1): 3–24. https://doi.org/10.1215/22011919-10216129

[26] Dougherty, M.L. 2016. From Global Peripheries to the Earth's Core The New Extraction in Latin America In: Deonandan, K. and Dougherty, M.L. (Eds), *Mining in Latin America: Critical approaches to the new extraction*. Routledge, 5.

[27] Veltmeyer, H. 2022. Extractivism and Beyond: Latin America debates. *The Extractive Industries and Society*, 11. https://doi.org/10.1016/j.exis.2022.101132;.Kingstone, P. 2018. *The Political Economy of Latin America: Reflections on Neoliberalism and Development after the Commodity Boom*. Taylor & Francis Group. 100.

[28] Gudynas, E. 2019. Value, Growth, Development: South American Lessons for a New Ecopolitics. *Capitalism Nature Socialism*. https://doi.org/10.1080/10455752.2017.1372502

the territorial expansion of extractive activities.[29] She situates this concept within the context of the 'commodities consensus' that emerged following the era of the Washington Consensus.

Within the energy transition literature, Fernandes argues that the growing demand for electric vehicles exemplifies how the discourse of the 'energy transition' operates to normalise extractivism in the Global South for the benefit of the Global North.[30] A further development of this broader debate may require engaging with the role of regional legal institutions in shaping, legitimising and replicating neo-extractivist development models across LATAM within the context of the energy transition and rising demand of minerals and metals. An initial approximation can be found in the work of Engstrom and Pérez, who explored the tensions between human rights, extractive projects and local communities with a focus on indigenous peoples.[31]

6.3.2 Extractivism in Peru

An essential characteristic of Peru's copper mining sector is its high export dependence on unrefined copper ores and concentrates.[32] This suggests that while Peru is a major producer, its domestic capacity for processing and refining copper is somewhat limited, meaning the country may not be fully capturing the potential added value from its copper resources.[33] Despite this, the export of minerals, particularly copper ore, represented some 9.5% of the country's GDP and 21% of its total exports by value in

[29] Svampa, M. 2015. Commodities Consensus: Neoextractivism and Enclosure of the Commons in Latin America. *South Atlantic Quarterly*, (1): 65–67. https://doi.org/10.1215/00382876-2831290

[30] Fernandes, S. 2024. Just Means 'Just' Everywhere: How Extractivism Stands in the Way of an Internationalist Paradigm for Just Transitions. *International Journal of Politics, Culture, and Society*, *37*(4): 493–511. https://doi.org/10.1007/s10767-024-09475-4

[31] Engstrom, P. and Pérez, E. 2025. Confronting Extractivism. In: Behrend, J. and Whitehead, L. (Eds). *Mounting Pressures on the Rule of Law*. Routledge.

[32] World Bank. 2021. *Peru - Mining Sector Diagnostic* (English). Available at: http://documents.worldbank.org/curated/en/448141632473923087 (Accessed: 8 August 2025).

[33] Bamber, P. *et al.* 2023. Innovation and Competitiveness in the Copper-Mining GVC: Developing Local Suppliers in Peru. *Industrial and Corporate Change*, *33*(4): 940, 941. https://doi.org/10.1093/icc/dtad033

2023.[34] The bulk of these copper ore exports, about 78% in 2023 were sold to China, and to a lesser degree to Japan (6.5%) and South Korea (3.7%).[35] Beyond the direct contributions from extraction and processing, the mining sector also generates considerable indirect economic activity through related industries and services such as transportation, construction, and supply chains.[36]

Of relevance for enabling and materialising this extractivism ecosystem were the legal reforms propelled during the 1990s, under the administration of President Alberto Fujimori. Peru implemented a series of neoliberal reforms known as 'shock therapy.'[37] These encompassed the privatisation of strategic economic sectors, incentivising trade aimed at an export-led economy, attracting foreign direct investment, and the establishment of tax stability regimes for mining companies.[38] These measures were also accompanied by the employment of military forces for dismantling paramilitary groups. Additionally, Peru entered into various international trade and investment agreements while fully endorsing the Washington Consensus.[39] As a result, the Peruvian mining sector expanded significantly,[40] becoming one of the most prominent contributors to the national economy and representing a substantial proportion of the country's exports.[41] The 2000s witnessed a new commodities boom, driven by the

[34] The Atlas of Economic Complexity. 2025. Peru. Available at: https://atlas.hks.harvard.edu/countries/604 (Accessed: 25 June 2025).

[35] The Atlas of Economic Complexity. 2025. Copper Ore in Peru. Available at Copper Ore in Peru Trade |The Observatory of Economic Complexity (Accessed: 8 August 2025).

[36] Sabatini, C. and Binetti, B. 2024. State Capacity, Mining and Community Relations in Peru: How Sustained Multi-Sectoral Dialogue Can Help Secure the Country's Role in Global Supply Chains. Chatam House.

[37] Kamphuis, C. 2012. Foreign Mining, Law and the Privatization of Property: A Case Study from Peru. *Journal of Human Rights and the Environment*, *3*(2): 217, 222.

[38] Bury, J. 2005. Mining Mountains: Neoliberalism, Land Tenure, Livelihoods, and the New Peruvian Mining Industry in Cajamarca. *Environment and Planning A: Economy and Space*, *37*(2): 221, 224. https://doi.org/10.1068/a371

[39] Crabtree, J. 2024. Peru and the Search for a Development Model That Works. In: Amann, E. and Figueiredo, P.A. (Eds). *Innovation, Competitiveness, and Development in Latin America*. Oxford University Press New York, 394.

[40] Li, F. 2015. *Unearthing Conflict: Corporate Mining, Activism, and Expertise in Peru*. Duke University Press, 15–20.

[41] World Bank. 2021. Peru - Mining Sector Diagnostic (English). Washington, D.C. Available at: http://documents.worldbank.org/curated/en/448141632473923087 (Accessed: 8 August 2025).

rapid industrialisation of China and India,[42] which in turn led to a marked increase in foreign mining investment within Peru.[43]

While copper mining has undeniably fuelled Peru's economic growth, it is essential to acknowledge the significant environmental and social considerations that accompany this activity. The extraction and processing of copper ore can cause a range of negative environmental impacts, including water contamination from mine tailings, acid mine drainage, and the release of heavy metals into rivers and groundwater.[44] Air pollution, caused by dust, gaseous emissions like sulphur dioxide, and particulate matter from smelting operations, is also a major concern in mining areas.[45] Furthermore, soil contamination with heavy metals and chemicals can render land unusable and pose risks to human and ecological health.[46] Deforestation and habitat loss are often associated with the expansion of mining operations, impacting biodiversity. The sheer volume of mine tailings and waste rock generated requires careful and responsible management to prevent long-term environmental damage.[47]

For many communities in Peru the industry provides essential employment and drives local economic activity. This dependence can be particularly pronounced in regions with limited alternative economic opportunities. It may result in human rights trade-offs, such as the right to health and to live in a clean environment versus the right to work.[48] Such

[42] Spalding, R.J. 2023. *Breaking Ground: From Extraction Booms to Mining Bans in Latin America*. Oxford University Press, New York, 8.

[43] Dargent, E. *et al.* 2017. Greater State Capacity, Lesser Stateness: Lessons from the Peruvian Commodity Boom. *Politics & Society*, *45*(1). https://doi.org/10.1177/0032329216683164

[44] Cacciuttolo, C. *et al.* 2023. Socio-Environmental Risks Linked with Mine Tailings Chemical Composition: Promoting Responsible and Safe Mine Tailings Management Considering Copper and Gold Mining Experiences from Chile and Peru. *Toxics*, *11*(5). https://doi.org/10.3390/toxics11050462

[45] Ibid.

[46] Dudka, S. and Adriano, D.C. 1997. Environmental Impacts of Metal Ore Mining and Processing: A Review. *Journal of Environmental Quality*, *26*(3): 590–602. https://doi.org/10.2134/jeq1997.00472425002600030003x

[47] Cacciuttolo, C. *et al.* 2023. Socio-Environmental Risks Linked with Mine Tailings Chemical Composition: Promoting Responsible and Safe Mine Tailings Management Considering Copper and Gold Mining Experiences from Chile and Peru. *Toxics*, *11*(5). https://doi.org/10.3390/toxics11050462

[48] Valencia, A. 2014. Human Rights Trade-Offs in a Context of 'Systemic Lack of Freedom': The Case of the Smelter Town of La Oroya, Peru. *Journal of Human Rights*, *13*(4). https://doi.org/10.1080/14754835.2014.960962

reliance also creates vulnerabilities. Over-dependence on a single industry can leave communities economically fragile, especially in the face of fluctuating global commodity prices or mine closures due to depletion or other factors. Mine closures can lead to significant job losses and economic hardship for communities that have become heavily reliant on mining as their primary source of income.[49] These are structural issues arising from extractivism models of development, which remain crucial to consider both nationally and internationally when governing the energy transition, as the demand for minerals and metals is set to increase dramatically.

6.4 Towards 'Just' Mining? La Oroya and the Decision of the Inter-American Court of Human Rights

La Oroya is a town located in the Peruvian Andes, recognised as one of the most heavily polluted sites on the planet because of the operation of a metallurgical complex since the 1920s.[50] It has been designated a 'sacrifice zone'. Recent empirical evidence indicates that for most of the year, residents of La Oroya are exposed to airborne toxic substances that exceed the safety thresholds established by the World Health Organization (WHO).[51] The WHO's recommended maximum daily limit for sulphur dioxide (SO_2) exposure of 40 $\mu g\ m^{-3}$ was exceeded on 610 days of the study period (2018–2020).[52] Exposure to high concentrations of this gas can lead to respiratory and lung diseases, with children being the most vulnerable due to the smaller size and ongoing development of their organs.[53] Alarmingly, children living close to mining operations in La Oroya exhibited elevated concentrations of heavy metals in their bodies, which can

[49] Neumann, P. 2016. Toxic Talk and Collective (In)Action in a Company Town: The Case of La Oroya, Peru. *Social Problems*, *63*(3): 431–446. https://doi.org/10.1093/socpro/spw010

[50] Li, F. 2015. *Unearthing Conflict: Corporate Mining, Activism, and Expertise in Peru*. Duke University Press, 35.

[51] Espinoza-Guillen, J.A. *et al.* 2023. Immission Levels and Identification of Sulfur Dioxide Sources in La Oroya City, Peruvian Andes. *Environment, Development and Sustainability*, *25*: 12843–2872. https://doi.org/10.1007/s10668-022-02592-0

[52] Ibid.

[53] Espinoza-Guillen, J.A. *et al.* 2023. Immission Levels and Identification of Sulfur Dioxide Sources in La Oroya City, Peruvian Andes. *Environment, Development and Sustainability*, *25:* 12843–12872. https://doi.org/10.1007/s10668-022-02592-0

lead to numerous and severe health complications, including cancer, respiratory, neurological and kidney diseases.[54]

In the *La Oroya case*, the IACHR ordered, along with individual compensation for the victims,[55] that the Peruvian state should investigate and identify those responsible for the threats and harassment experienced by environmental defenders, as well as those accountable for the environmental degradation resulting from the contamination in La Oroya.[56] Furthermore, the victims are entitled to receive appropriate medical treatment.[57] Regarding environmental harm, the IACHR ordered the State of Peru to develop a remediation plan, which must include baseline assessments of contamination levels in air, water, and soil.[58]

6.4.1 Restorative Justice and the Decision of the Inter-American Court of Human Rights

While the IACHR's decision in the La Oroya case represents a milestone in mobilising International Human Rights Law (IHR) to tackle environmental problems, it can also be critiqued from the perspective of restorative justice due to the inherent nature of international adjudication as a mode of an international dispute resolution solution. IHR adjudication may constitute an inadequate mechanism for advancing environmental protection and mitigating the adverse health impacts arising from environmental degradation. A first reason is that obligations under international law are obliged exclusively to states and are constrained by territorial jurisdiction.[59] Additionally, responsibility must meet the requirement of

[54] Piñeiro, X. *et al.* 2021. Heavy Metal Contamination in Peru: Implications on Children's Health. *Nature Scientific Reports, 11*, 22729. https://doi.org/10.1038/s41598-021-02163-9

[55] Case of La Oroya Population v. Peru. (Judgment of 27 November 2023, on Preliminary Objections, Merits, Reparations and Costs) Inter-American Court of Human Rights Ser C No 511 [366]–[378].

[56] Ibid.

[57] Ibid.

[58] Ibid.

[59] Bellinkx, V. *et al.* 2022. Addressing Climate Change through International Human Rights Law: From (Extra)Territoriality to Common Concern of Humankind. *Transnational Environmental Law, 11*(1): 69–93. doi:10.1017/S204710252100011X; Min. Z and Wimer, F. 2025.Transnational Governance of Soybean Land Use in South America: A Polycentric Approach. *Transnational Environmental Law, 14*(1):145–170. https://doi.org/10.1017/S2047102524000268

a causal link between the human rights violation to an identified group of individual victims and a specific act or omission attributable to the state. Consequently, within the fora of international human rights adjudication, it is impracticable to directly hold multinational corporations accountable for environmental harm. Even though these actors are responsible for taking decisions and have the power to implement strategies that result in environmental harm and human health problems.

Related to this reason is the issue of power imbalances between states and non-state actors at the international level. The employment of IHR law is furthermore constrained because some states, especially from the Global South, remain structurally disadvantaged in terms of power relations when compared to large multinational enterprises. This imbalance affects not only the formulation of environmental quality regulations but also their effective enforcement. At the time of writing this piece, the Peruvian government enacted legislation aimed at suppressing the legal personality of non-governmental organisations that challenge governmental actions.[60] This measure appears to have been prompted by the government in the immediate aftermath of the recent La Oroya judgement issued by the IACHR. Another example is how the land rights of the Negritos community were rescinded to benefit the multinational corporation Yanacocha in its gold mining operations.[61]

Such situations are worsened by the fact that states in the Global South often exhibit institutional barriers which makes it even more difficult for local communities access to justice, precisely where many multinational corporations operate and maintain subsidiaries. This dynamic has prompted environmental activists to mobilise and pursue human rights litigation in jurisdictions within the Global North, where numerous transnational corporations are registered. Notable examples include the Monterrico case[62] and the BHP case.[63] While IHR adjudication presents a significant

[60] Office of the High Commissioner for Human Rights. 2025. Peru: Impact of international cooperation law on NGOs. Available at: https://www.ohchr.org/en/press-releases/2025/04/peru-impact-international-cooperation-law-ngos (Accessed: 25 June 2025).

[61] Kamphuis, C. 2012. Foreign Mining, Law and the Privatization of Property: A Case Study from Peru. *Journal of Human Rights and the Environment*, *3*(2): 217–253.

[62] Lindt, A. 2020. Transnational Human Rights Litigation: A Means of Obtaining Effective Remedy Abroad?. *Journal of Legal Anthropology*. https://doi.org/10.3167/jla.2020.040204

[63] Farrington, F. 2021. Municipio de Mariana v BHP Group: Implications of the UK High Court's Decision. *Business and Human Rights Journal*, *6*(2): 392–398. doi:10.1017/bhj.2021.10

opportunity to promote corporate accountability within national jurisdictions, it maintains existing power imbalances both within and between countries as well as between countries and multinational corporations.

In the *La Oroya case*, the measures adopted by the Court were directed solely at the State of Peru, which undoubtedly constitutes an approach consistent with established principles of International Law of territorial sovereignty. This underscores the limitations of international law as a mechanism for advancing a just transition, particularly in contexts where industrial operations involve multinational corporations and weak states. The metallurgical complex located in La Oroya includes such corporations, which are fully cognisant of the hazardous chemicals and toxic substances produced because of their profit-driven industrial activities. Indeed, the decision is grounded in Peru's responsibility in the violation of the preventive principle,[64] which subsequently necessitated restorative measures to address the environmental harm. But as mentioned above, the principle of restorative justice emphasises that responsibility is central to addressing harm and fostering healing. Specifically, it holds that offenders who have caused harm should take responsibility for their actions by acknowledging the harm they have caused. In the La Oroya case the state of Peru, not the actual offenders, namely the transnational companies, was held responsible.

However, this remediation is highly dependent on the state's legal enforcement of the international court's decisions and the availability of resources. On 5 February 2025, the IACHR found that Peru had only partially complied with the judgment.[65] The state still needs to complete investigations into the contamination in La Oroya and provide adequate reparations to the victims. To prevent such damage from arising, a just transition framework informed by a restorative justice perspective should prioritise interventions at the early stages of mining activities, that is, before exploration activities, which includes active engagement with local communities and sharing information about probable environmental

[64] Case of La Oroya Population v. Peru. (Judgment of 27 November 2023, on Preliminary Objections, Merits, Reparations and Costs) Inter-American Court of Human Rights Ser C No 511 [332].

[65] Case of La Oroya Population v. Peru. (Judgment of 27 November 2023, on Preliminary Objections, Merits, Reparations and Costs) Inter-American Court of Human Rights Ser C No 511 'Supervisión de Cumplimiento de Sentencia. Resolución de la Corte Interamericana de Derechos Humanos.' February 3, 2025.

impacts. A combination between domestic law[66] and international law[67] should provide the necessary regulations and standards to make feasible this approach. International adjudication in this sense should be the last resort to obtain reparations for both the victims and the damaged environment through an adequate restoration plan.

A second reason pertains to the adversarial approach employed in resolving IHR disputes, which fails to account for the structural conditions that enable the persistent violation of human rights. As argued by Valencia[68] the victim-perpetrator-remedy model[69] is inadequate in capturing the complex root causes of the persistent human rights violations experienced by the inhabitants of La Oroya. This is because the structural injustices underlying the case do not cease when the State of Peru and the Doe Run company acknowledge responsibility and formulate a remedial plan for the victims, but when the '*causes of causes*' are properly addressed. Such causes are related to the historical, social, economic and political conditions that provide the conditions for an unequal distribution of social and environmental burdens between and within countries. In La Oroya, these issues are linked to historical dependencies on extractive industries, reinforced by neoliberal policies and global institutions, that can be framed as an extractivism model of development. Recognising such conditions reinforces the importance of the principle of restorative justice, as it directs attention to who should bear responsibility for restoring both

[66] See Caroline Omari Lichuma, '(Laws) Made in the 'First World': A TWAIL Critique of the Use of Domestic Legislation to Extraterritorially Regulate Global Value Chains' (2021) 81 Zeitschrift für ausländisches öffentliches Recht und Völkerrecht / Heidelberg Journal of International Law, 497.

[67] Perrone, N.M. 2025. Mission Impossible? Promoting Global Sustainable Development in Times of National Industrial Policies. *Global Policy*. Available at: https://www.globalpolicyjournal.com/articles/climate-change-energy-and-sustainability/early-view-article-mission-impossible-promoting

[68] Valencia, A. 2014. Human Rights Trade-Offs in a Context of 'Systemic Lack of Freedom': The Case of the Smelter Town of La Oroya, Peru. *Journal of Human Rights*, *13*(4): 456–479. https://doi.org/10.1080/14754835.2014.960962

[69] The victim-perpetrator-remedy model is based on assigning responsibility to the perpetrator for their actions or omissions. As such, it emphasises in identifying the concrete damage suffered by the victim because of the actions or omissions of the perpetrator rather than addressing the structural conditions that enabled such a situation to occur. Valencia, A. 2014. Human Rights Trade-Offs in a Context of 'Systemic Lack of Freedom': The Case of the Smelter Town of La Oroya, Peru. *Journal of Human Rights*, *13*(4): 456–479. https://doi.org/10.1080/14754835.2014.960962

the environmental conditions that existed prior to the mining project and the human relationships surrounding it.

Legal scholars argue at this respect that IHR facilitates extractivism, dispossession, and the externalisation of environmental harms from the Global North to the Global South.[70] Public International Law has traditionally been rooted in Eurocentric doctrines shaped by Enlightenment thought, which emphasise a dichotomy between humanity and nature,[71] thereby naturalising human domination over the environment. IHR law is likewise embedded within this conceptual framework. While it offers strategic advantages for environmental protection in legal cultures where the right to live in a healthy environment is neither internationally nor constitutionally recognised,[72] it is equally important to acknowledge its structural limitations. An example is given in the postcolonial era – particularly during the 1970s in Latin America – when Global North states frequently invoked human rights to protect the foreign investments of their corporations during nationalisation processes involving strategic natural resources.[73] A restorative justice approach may challenge these circumstances where IHR is rooted, as it requires the engagement of all stakeholders across the global value chain of mining for the energy transition. In this sense, measures to address socio-environmental impacts caused by mining should also consider a bottom-up approach. Lotter suggests in this respect a restorative justice should look at the actors involved, '*the mining industry as the 'perpetrator', and the environment and communities as the 'victim' of harm.*'[74] Thus, further innovative ways apart from international adjudication in international law for engaging and obliging relevant actors that participate in the mining of Global Value Chains should be considered within the just transition debate.

[70] Gonzalez, C. 2015. Environmental Justice, Human Rights, and the Global South. *Santa Clara Journal of International Law*, *13*(1). https://digitalcommons.law.scu.edu/scujil/vol13/iss1/8

[71] Ibid.

[72] The UN General Assembly (UNGA) adopted resolution A/RES/76/300 on the right to a healthy environment which serves as a springboard for policies to protect people and nature.

[73] Natarajan, U. 2022. Who Do We Think We Are?: Human Rights in a Time of Ecological Change In: Natarajan, U. and Dehm, J. (Eds) *Locating Nature:Making and Unmaking International Law*. Cambridge University Press, 207.

[74] Lotter, A. 2020. Part 2: Mine(d) Over Matter: Restorative Justice in the Mining Industry. Available at: https://law.uct.ac.za/mineral-law/articles/2020-02-05-part-2-mined-over-matter-restorative-justice-mining-industry (Accessed: 25 June 2025).

IHR adjudication also fails to adequately capture the complex and multi-scalar dimensions involved in mining for the energy transition. The impacts of mineral extraction for this purpose must be assessed not only at national and local levels but also from a global perspective. In this regard, it has been argued that mining law for the energy transition should integrate not only domestic and subnational normative systems but also global legal and regulatory frameworks.[75] The mining industry operates within intricate global value chains, involving a multitude of actors from diverse geographical contexts who contribute labour and resources towards the production of goods essential to the energy transition. These normative structures were significantly reshaped during the neoliberal era to facilitate international trade; as such, they must now be critically re-evaluated and reformed to allow for a holistic assessment of fairness and equity across the various actors and regions implicated in global value chains. A restorative justice approach, thus, means that norms applying to mining extraction should not be assessed only at the local and national level but also considering the lifecycle of the product and/or global value chains involved.

Another pitfall of the IACHR's decision in the La Oroya case is its ignorance of existing dependencies. Dominant thought in political economy explores the underdevelopment of countries in the Global South and developed the 'Dependency Theory'. The theory holds that the global economic system is structured to perpetuate inequality between wealthy, industrialised 'core' nations and poorer, resource-exporting 'peripheral' nations.[76] Peripheral countries are locked into a dependent relationship with core nations, exporting raw materials while importing manufactured goods, which prevents them from developing diversified, self-sustaining economies. Such dependency, rooted in colonialism and reinforced by global trade dynamics, creates a cycle of underdevelopment, leaving peripheral nations vulnerable to price fluctuations and exploitation by multinational corporations and international financial institutions.

Looking at the history of the case of La Oroya itself, a form of this economic dependency becomes strikingly visible. Despite that, in April 2002, a group of residents and non-governmental organisations officially

[75] Bastida, A.E. 2020. *The Law and Governance of Mining and Minerals: A Global Perspective*. Hart Publishing, 5.

[76] Prebisch, R. 1950. The Economic Development of Latin America and Its Principal Problems. United Nations Department of Economic Affairs, Economic Commission for Latin America (ECLA), New York.

formed the MOSAO (Movement for the Health of La Oroya) to address the health and environmental problems facing the community; Neumann holds that such mobilisation proved challenging.[77] Locals were not easily convinced by the cause, and participants were even repeatedly threatened and physically attacked, often by unionised workers of the mine. She posits that La Oroya highlights the broader issue of how affected communities, even in cases of extreme environmental harm, struggle to advocate for change when their economic survival is tied to the very industries causing the damage. In addition, Antunes de Oliveira recently argued that both market-driven and state-led approaches to development have failed to address the structural issues of dependency and economic vulnerability.[78] Thus, even if the Peruvian state is ordered to undertake restoration by the IACHR, it might not possess the political will nor the institutional capacity[79] to prevent subsequent human rights violations. A restorative justice approach could help address these dependencies and, for instance, lay the groundwork for the establishment of binding responsibilities for transnational corporations.

6.5 Conclusion

While Latin America plays a critical role in supplying minerals for the global energy transition, the region's extractivist development model exhibits socio-environmental and human rights challenges. First legal attempts to ensure more sustainable and just mining of materials have been undertaken, but by analysing the IACHR decision of the La Oroya case, it can be demonstrated that there are limitations of existing legal frameworks in addressing structural injustices and ensuring corporate accountability. While the IACHR decision in *Inhabitants of La Oroya v. Peru* represents significant progress in environmental justice in LATAM, it simultaneously contributes to sustaining the structural conditions that allow for environmental harm and unjust working conditions in LATAM. Peru is economically dependent on raw material exports, a system rooted

[77] Neumann, P. 2016. Toxic Talk and Collective (In)Action in a Company Town: The Case of La Oroya, Peru. *Social Problems*, *63*(3) 431–446. https://www.jstor.org/stable/44015389

[78] de Oliveira, F.A. 2024. *Dependency and Crisis in Brazil and Argentina: A Critique of Market and State Utopias*. University of Pittsburgh Press.

[79] Crabtree, J. and Durand, F. 2017. *Peru: Elite Power and Political Capture*. Zed Books.

in colonial histories and reinforced by global trade dynamics, which traps the state in underdevelopment while they bear the environmental and social costs. neoliberal policies, such as those introduced in the 1990s in Peru, and political capture further weakened the state by prioritising foreign investment through tax incentives, privatisation, and deregulation and enabling transnational corporations to operate with minimal accountability. In these conditions, the IACHR decision, which sorely holds the state of Peru but not the transnational corporations accountable, is incapable of creating real justice on the ground. The La Oroya case underscores the urgent need for a fundamental shift in prevailing development paradigms. Moving beyond extractivism models towards more sustainable and equitable alternatives is essential in order to prioritise environmental protection and the well-being of affected communities.

Building on these insights, we advocate for a restorative justice approach as a transformative framework to achieve a just and sustainable transition in the mining sector. A restorative justice emphasis within the framework of a just transition offers a transformative perspective, as it facilitates the repair of fractured relationships in contexts of socio-environmental conflict while also creating avenues for the prevention of environmental harm. The potential of restorative justice approaches lies in their capacity to prevent, manage, and remediate environmental degradation and related social conflicts. Embedding this dimension of just transition into international, regional, and national legal frameworks could contribute to the early prevention of environmental damage by establishing binding responsibilities for transnational corporations at the initial stages of mining projects. Simultaneously, it would reinforce the obligation of states to ensure the protection of human rights throughout the extractive processes. This way, it would not only be possible to move towards more 'just' mining practices in LATAM but also create possibilities for those countries to reap the benefits of their abundance of minerals and metals for the global energy transition.

Interestingly, a path to look forward is to dig deeper into the recent advisory opinion issued by the IACHR (AO-32/25) regarding the state's obligations on the climate emergency. According to paragraph 342, the Court emphasises the need to protect the human rights of those affected by mining projects aimed at extracting minerals for the energy transition, including protections of labour rights and those affected by relocation as a consequence of these projects.

PART IV

Reduce: Lower Greenhouse Gas Emissions

CHAPTER 7

A Procedural Justice Analysis of Cross-border Green Hydrogen Initiatives Between the EU and Tunisia

Paola Jimenez-Casanova, Shakya A. Wickramanayake, Sonia Chabane, Strahinja Obrenović, and Syed Sajad Ali Shah

Abstract The EU's hydrogen strategy, which aims to develop a European hydrogen market to decarbonise energy-intensive industries that are hard to electrify, relies partly on imports of green hydrogen from third countries such as Tunisia. Tunisia alone could potentially supply up to 40% of the green hydrogen target established in the REPowerEU Plan. The SoutH2 Corridor

P. Jimenez-Casanova
Centre for Environmental Law Studies (CEDAT), Universitat Rovira i Virgili (URV), Tarragona, Spain
e-mail:paola.jimenez@urv.cat

S. A. Wickramanayake
Tilburg Institute of Law, Technology and Society (TILT), Tilburg, The Netherlands

S. Chabane
Political Science Department, Université Libre de Bruxelles (ULB), Brussels, Belgium

S. Obrenović
University of Belgrade, Belgrade, Serbia

S. Sajad Ali Shah
UNU-EHS, University of Bonn, Bonn, Germany

A. Phillips, L. Kaschny (eds.), *Global Perspectives on the Circular Economy in the Just Transition*, Just Transitions,
https://doi.org/10.1007/978-3-032-18342-2_7

– a cross-border infrastructure project included in the Union's list of Projects of Common Interest and Projects of Mutual Interest (PCIs and PMIs) – seeks to connect Tunisia with the European Union (EU) through a pipeline extending from Italy to Austria and Germany, with the goal of supplying the latter two with green hydrogen by 2030. This chapter examines the risks that such projects may pose for ensuring an inclusive and equitable energy transition in Tunisia. Given Europe's historical relationship with Tunisia, including through colonisation, there are concerns that these energy projects will potentially perpetuate historical patterns of dominance, dependency, and extractivism. By analysing the challenges facing the EU, EU Member States and the Tunisian energy sectors in pursuing a just energy transition, alongside Tunisia's democratic backsliding, the chapter highlights the absence of adequate procedural justice safeguards in some of the public and private legal arrangements underpinning the SoutH2 Corridor project. As such, it appears that said gaps strongly risk undermining the just energy transition, especially in Tunisia. The chapter concludes that the EU bears a responsibility to ensure that a set of minimum safeguards are established to prevent the reproduction of colonial practices in its external energy relations, using Tunisia as an illustrative example of broader risks faced by other partner countries in similar contexts.

Keywords European Union • Tunisia • Green Hydrogen • Energy justice • Procedural justice • Energy Transition • Green Deal • Global Gateway

7.1 Introduction

The EU's energy sector is undergoing a transition to become carbon neutral by 2050.[1] The decarbonisation of the energy sector is considered necessary not only to mitigate climate change but also to increase energy security in the European Union (EU) and increase the competitiveness of its industries.[2] The EU is implementing a series of policies to

[1] As part of the European Green Deal, the European Union (EU)'s roadmap to sustainability and achieve the Union's climate neutrality target by 2050. The 2050 target was made legally binding through the European Climate Law, alongside the interim target of at least 55% greenhouse gas emission reduction by 2030. European Commission. 2019. Communication COM(2019) 640 final, The European Green Deal, (EU Green Deal); European Commission. 2021. Communication COM(2021) 550 final, Fit for 55': delivering the EU's 2030 Climate Targets on the way to climate neutrality.

[2] European Commission. 2022. Communication COM(2022) 230 final, REPowerEU Plan (REPowerEU Plan); European Commission. 2025. Communication COM(2025) 85

become carbon neutral. These include increasing direct electrification of end-consumption, as well as the creation and expansion of a renewable and low-carbon hydrogen market targeting the so-called hard-to-decarbonise sectors, where direct electrification is unfeasible.[3] Initially, the EU established a target of 10 million tonne renewable hydrogen production by 2030, which was meant to substitute the consumption of fossil fuel-based hydrogen in the EU.[4] However, this target was updated and doubled in the REPowerEU Plan to accelerate the uptake of the hydrogen market, with a renewable hydrogen production and import target of 10 million tonnes each by 2030.[5]

Against this background, there is an initiative to develop a purely dedicated hydrogen system, the EU Hydrogen Backbone (hereafter, 'EU H2 Backbone')[6] which set out a vision for a pan-European hydrogen network.[7] For its development, the EU and its Member States have been entering into partnerships with third countries to support the production

final, The Clean Industrial Deal: A joint roadmap for competitiveness and decarbonisation (Clean Industrial Deal); and European Commission. 2025. Communication COM(2025) 79 final, Action Plan for Affordable Energy.

[3] European Commission. 2020. European Commission. 2022. Communication COM-Communication COM(2020) 301 final, a hydrogen strategy for climate-neutral Europe (EU Hydrogen Strategy); European Commission. 2020. Communication COM(2020) 299 final, Powering a climate-neutral economy: An EU Strategy for Energy System Integration (EU Energy System Integration Strategy).

[4] According to a Clean Hydrogen Partnership report, renewable hydrogen production through water electrolysis accounted for only 0.4% in 2023. Clean Hydrogen Partnership, 'The European hydrogen market landscape,' November 2024. Available at https://observatory.clean-hydrogen.europa.eu/sites/default/files/2024-11/The%20European%20hydrogen%20market%20landscape_November%202024.pdf (Accessed: 11 August 2025).

[5] REPowerEU Plan; For a description of the political context and energy policies shaping hydrogen regulation in the EU see Jiménez Casanova, P. 2024. Towards a Sustainable, Integrated, and Decarbonized Energy System in the EU: Addressing Structural Challenges Through Hydrogen System Planning. *Journal for European Environmental & Planning Law*, *21*(3–4), 193-214. https://doi.org/10.1163/18760104-21030004 The Commission has announced the revision of the Hydrogen Strategy in 2026, this may lead to a modification of the current targets to a lower production and import target. Available at: https://www.hydrogeninsight.com/policy/eu-to-discuss-potential-changes-to-rfnbo-rules-as-part-of-2026-revision-of-blocs-hydrogen-strategy/2-1-1882918 (Accessed: 17 October 2025).

[6] European Hydrogen Backbone. 2022. Five Hydrogen Supply Corridors for Europe in 2030. Available at: https://ehb.eu/files/downloads/EHB-Supply-corridor-presentation-Full-version.pdf/ (Accessed:13 August, 2025).

[7] Ibid.

and import of renewable electricity and green hydrogen into the EU.[8] Due to their geographical proximity and existing trade relationships, North African countries, such as Morocco, Algeria, Egypt and Tunisia, have the potential to become relevant trading partners with the EU in the transition to decarbonised energy systems, especially through the trade of renewable hydrogen and electricity by 2030 and beyond.

The EU's flagship development cooperation strategy, the Global Gateway, aims to facilitate such partnerships.[9] However, due to Europe's colonial heritage, characterised by resource extractivism,[10] dependency and power imbalances, there are well-grounded concerns that they may reproduce forms of colonialism in the present, through specific projects and practices, to the disadvantage of formerly colonised countries,[11] such as modern Tunisia, Algeria, Egypt and Morocco. Green hydrogen projects, in particular, have been criticised for being instruments of the EU's 'green colonialism' in North Africa.[12] This critique stands in contrast to the EU's claim that its green hydrogen initiatives under the Global Gateway (hereafter 'H2GG projects') are sustainable and mutually beneficial for all partners and will ultimately empower local communities in third countries.[13]

As many of these projects are still in the initial planning or development phase, it is not yet possible to investigate whether the H2GG projects will

[8] Furthermore, the EU's Carbon Border Adjustment Mechanism (CBAM) includes grey hydrogen and derivatives, creating higher incentives for third countries for developing green hydrogen projects to supply the EU markets.

[9] The Global Gateway strategy was launched by the European Commission in December 2021, with the aim to promote projects and infrastructure in partner countries, especially to tackle climate change. See European Commission. 2021. Joint Communication, JOIN(2021) 30 final, The Global Gateway (The Global Gateway).

[10] Extractivist scholars, coming from diverse academic and activist traditions in Latin America, have enabled the emergence of this concept in the 1990s (*extractivismo*). While there is no consensus on the boundaries of extractivism and its underlying theoretical framework, as an object of study, in this instance, it broadly encompasses the 'expansion of resource-frontiers and the increasing social conflict between extractive companies and local communities'. See: Malena Antmann. 2025. Capitalism and resource extraction: A Marxist value-theory approach to Latin American socio-environmental conflicts. *The Extractive Industries and Society, 22*. https://doi.org/10.1016/j.exis.2025.101617

[11] Dejonghe, M. and Van de Graaf, T. 2025. Green colonialism or green transformation? The equity implications of clean hydrogen trade. *Political Geography*, 120. https://doi.org/10.1016/j.polgeo.2025.103338

[12] Hamouchene, H. and Sandwell, K. (eds). 2023. *Dismantling Green Colonialism: Energy and Climate Justice in the Arab Region*. Pluto Press.

[13] European Commission. 2021. Joint Communication, JOIN (2021) 30 final, The Global Gateway.

exploit third countries and former colonies for the benefit of the EU, nor the extent of such exploitation. However, given that the EU claims that these projects will be mutually beneficial, the onus should be on the EU to prove that such scenarios would not arise. In particular, to demonstrate that all H2GG projects are not extensions of colonialism in the present and are instead facilitating a just transition. The threat of such exploitation is particularly heightened if third countries themselves have strained energy systems and are politically unstable or undemocratic, as is the case of Tunisia.

Nonetheless, the North Africa and Southern Europe Corridor of the EU H2 Backbone (hereafter, 'NAS Corridor'), a hydrogen pipeline corridor in the EU that aims to physically connect the EU (via the South of Italy) with the North of Africa (through Tunisia), offers us an opportunity to explore whether the H2GG projects can instead be part of a just energy supply chain – one where the burdens of producing green hydrogen are not disproportionately endured by third countries, and where benefits are not concentrated in the EU. An example of how these supply chains would be unfair is related to the environmental costs of building and operating the infrastructure of these green hydrogen systems. Primarily, negative externalities will be borne by third countries, whereas environmental benefits of using clean energy (reduction of carbon emissions) are enjoyed primarily by the EU. This is due to these projects prioritising the export of clean energy to the EU, over supplying the domestic markets. The NAS Corridor involves the development of more than 3,300 km of dedicated hydrogen pipelines that will run through Germany, Austria and Italy in the EU, extending up to Tunisia and Algeria, with the main objective to supply renewable hydrogen for consumption in Germany and Austria.[14] The corridor proposes to connect green hydrogen production in Tunisia and Algeria to EU markets. It essentially covers more than 40% of the REPowerEU import target, making it a vital element of EU energy security. The SoutH2 Corridor, a multi-project development, has been included in the list of Global Gateway flagship projects under the energy and climate category.[15]

[14] The SoutH_2 Corridor. Available at: https://www.south2corridor.net/ and https://www.sea-corridor.com/en/h2-initiatives (Accessed: 13 August 2025).

[15] General Secretariat of the Council. 2024. Draft proposal for a list of Global Gateway flagship projects for 2025 and for amending the lists of Global Gateway flagship projects for 2023 and 2024. Available at: https://data.consilium.europa.eu/doc/document/ST-15281-2024-INIT/en/pdf (Accessed:13 August 2025); The South H2 Corridor consists of sever-

To analyse the risks of the EU engaging in neocolonial practices akin to green colonialism, this chapter focuses on the positive and negative effects that the development of the NAS Corridor may have on Tunisia, centring procedural justice as a key framework. The central question raised is whether the EU has a duty to ensure that the H2GG projects in third countries, such as Tunisia, are 'energy-just' and if it can fulfil this responsibility by incorporating procedural energy justice safeguards in their Global Gateway Memorandums of Understanding (MoUs). Ultimately, how these can influence their incorporation in the H2GG project agreements.[16] Tunisia's strategic location makes it a compelling case study, as it is a key partner for the EU to develop the cross-border infrastructure needed to enable hydrogen imports through the NAS Corridor. This chapter will not explore the manner in which due diligence and corporate responsibility obligations are imposed on the H2GG project promoters, given the early stages of such projects and ongoing deregulatory pushbacks at the EU level on corporate social responsibility (on both the Corporate Sustainability Reporting Directive and the Corporate Sustainability Due Diligence Directive).[17]

To address this issue, this chapter will begin by explaining the type of hydrogen projects that the EU is prioritising – renewable and low-carbon hydrogen – and the relevance of the definition for receiving EU support for their development outside the EU. Then, the section explores counterclaims made by scholars, communities and activists

al projects, including the development of purely dedicated hydrogen pipelines in Germany, Austria, Italy, Tunisia and Alergia, as well as the installation of electrolysers for hydrogen production, and solar and wind production plants.

[16] Given that the projects considered in the NAS Corridor are still in the negotiating phase, particularly the sections pertaining the maritime cross-border pipeline and the hydrogen production plants in Tunisia, the scope of the article is limited to the international legal mechanisms that exist today and the rights and obligations contained in them.

[17] Spinaci, S. 2025. Omnibus I - sustainability reporting - proposal amending CSRD and CSDDD - in 'A new plan for Europe's sustainable prosperity and competitiveness'. European Parliament Legislative Train Schedule. Available at: https://www.europarl.europa.eu/legislative-train/package-simplification-business/file-first-omnibus-package-on-sustainability-proposal-amending-csrd-and-csddd (Accessed: 13 August 2025; Euractiv. 2025. MEPs endorse rollback of corporate sustainability rules on EPP terms. Available at: https://www.euractiv.com/news/meps-endorse-rollback-of-corporate-sustainability-rules-on-epp-terms/ (Accessed: 20 October 2025).

regarding green colonialism, and it describes local hydrogen policies in Tunisia (Section 7.2). The chapter then discusses why green hydrogen projects may pose more challenges than benefits for the local population and explain the need for safeguards to ensure the equitable distribution of benefits stemming from these extractive projects. The chapter then continues to analyse these counterclaims from the perspective of energy justice and establishes procedural safeguards that green hydrogen project developers need to consider to minimise the risk of engaging in neocolonial practices (Section 7.3). Thereafter, this chapter discusses the current political and socio-economic circumstances surrounding the green hydrogen projects in Tunisia and the impact these have on ensuring energy justice in the country (Section 7.4). This section ends by elaborating on the key argument of this chapter, namely why the onus is on EU institutions to take steps to institute safeguards to avoid neocolonial practices in the operation of the green hydrogen projects, and why incorporating procedural justice elements in the MoUs and agreements between the EU (including private companies) and third countries, in particular Tunisia, is an ideal starting point to ensure energy justice in the green hydrogen supply chain. The chapter concludes by enunciating some of the pitfalls of the projects and some policy recommendations (Section 7.5).

To enable such findings, this chapter used a doctrinal legal analysis.[18] Besides academic literature and a content analysis of institutional agreements (including at EU and national levels), the analysis is supported by policy and official project documentation and business cases produced by companies. Furthermore, the analysis made use of news articles and reports from non-governmental organisations (NGOs) and transnational energy networks in French and English across Europe and in Tunisia to complement prior sources and provide details about those respective projects and their domestic reception.

[18] Doctrinal legal analysis is a methodology used legal research, where the relevant legal sources such as legislative statues, treaties, case law, and parliamentary discussions as well as legal academic literature are collated and understood through interpretive analysis. This analysis can be based on historical, contextual or normative perspective. See: Taekema, S., and van der Burg, W. 2024. Chapter 1: Introduction: Methodology of legal research. Contextualising Legal Research (Edward Elgar Publishing).

7.2 EU Hydrogen Rules and Tunisian Green Hydrogen Projects

7.2.1 EU Rules for Renewable and Low-Carbon Hydrogen

Hydrogen can be produced from multiple sources, including fossil fuels and renewable energies. The whole range of ways in which it can be produced gave birth to the hydrogen colour spectrum.[19] For instance, green hydrogen is typically considered as the hydrogen produced through the electrolysis of water, where the electrolyser is powered by a renewable source, resulting in a minimum GHG footprint.[20] Grey hydrogen is hydrogen produced from the reforming of natural gas with a medium GHG footprint,[21] and blue hydrogen when GHG is captured and stored underground.[22] Whereas the hydrogen colour spectrum is helpful to understand the production process of hydrogen, it is ultimately the legal definition and the regulatory framework that is relevant for industries, investors and policymakers, as they provide certainty of what counts as sustainable ways of producing hydrogen.[23]

In this sense, the EU's long-term strategy for developing a hydrogen market is considered necessary to facilitate the decarbonisation of hard-to-abate industries. From the supply side, the strategy is prioritising the production of green hydrogen from the electrolysis of water, via electrolysers, and with the electricity stemming from renewable sources, such as wind and solar energy – what it calls 'renewable hydrogen'[24] This is in

[19] World Energy Council. National Hydrogen Strategies. 2021. Available at: https://www.worldenergy.org/assets/downloads/Working_Paper_-_National_Hydrogen_Strategies_-_September_2021.pdf (Accessed: 11 August 2025).

[20] World Energy Council. National Hydrogen Strategies. 2021. Available at: https://www.worldenergy.org/assets/downloads/Working_Paper_-_National_Hydrogen_Strategies_-_September_2021.pdf (Accessed: 11 August 2025).

[21] Ibid.

[22] Ibid.

[23] For a description and analysis of different certification schemes for hydrogen and their relevance see Gallegos, F. 2024. Bridging the Gap: Certification Schemes for Sustainable Hydrogen in Global Trade. *Journal for European Environmental & Planning Law*, *21* (3–4), 239–273. https://doi.org/10.1163/18760104-21030005

[24] EU Hydrogen Strategy; Recital 13 of the Preamble to the European Commission. 2024. Directive (EU) 2024/1788 on common rules for the internal markets for renewable gas, natural gas and hydrogen, amending Directive (EU) 2023/1791 and repealing Directive 2009/73/EC, OJ L 15.7.2024 (Renewable Gases and Hydrogen Directive). Available at: https://eur-lex.europa.eu/eli/dir/2024/1788/oj/eng (Accessed on: 20 October 2025).

part because the EU wants to increase the competitiveness of European electrolyser technologies, expecting investments of around EUR 180 and 470 billion by 2050.[25] Furthermore, this will help to create new jobs and foster economic growth as a result of this new industry.[26] Nonetheless, for the short and mid-terms, low-carbon hydrogen production is considered necessary to decarbonise existing fossil fuel-based hydrogen production, as well as to power some electrolysers with low-carbon electricity production.[27]

Under EU legislation, renewable hydrogen is included under the broader definition of renewable fuels of non-biological origin (RFNBOs) included in the Renewable Energy Directive (RED III).[28] RFNBOs are defined as 'liquid and gaseous fuels the energy content of which is derived from renewable sources other than biomass'.[29] RFNBOs will play an important role in the decarbonisation of the transport sector, including long-haul transport, maritime and aviation, considered as hard-to-abate sectors.[30] In this sense, hydrogen production via electrolysers powered by renewable electricity is what can be accounted for as RFNBOs when it complies with the requirements established in RED III.[31] These requirements include the geographical correlation, temporal correlation and additionality criteria established in Commission Delegated Regulation (EU) 2023/1184,[32] and the 70% greenhouse gas (GHG) emissions

[25] EU Hydrogen Strategy. Available at: https://eur-lex.europa.eu/eli/dir/2024/1788/oj/eng (Accessed on: 20 October 2025).

[26] Ibid.

[27] EU Hydrogen Strategy and Recital 13 of the Preamble to the Renewable Gases and Hydrogen Directive.

[28] European Parliament and Council. 2023. Directive (EU) 2023/2413 amending Directive (EU) 2018/2001, Regulation (EU) 2018/1999 and Directive 98/70/EC as regards the promotion of energy from renewable sources, and repealing Council Directive (EU) 2015/652, OJ L 31.10.2023 (RED III).

[29] Article 2(36), RED III.

[30] EU Hydrogen Strategy; European Parliament. 2023. Regulation (EU) 2023/2405 on ensuring a level playing field for sustainable air transport (ReFuelEU Aviation), OJ L 31.10.2023; European Parliament. 2023. Regulation (EU) 2023/on the use of renewable and low-carbon fuels in maritime transport, and amending Directive 2009/16/ECEU, OJ L 234, 22.9.2023 (FuelEU Maritime).

[31] Articles 27.1 (a) (ii), (d), 27.6, and 29a of RED III.

[32] European Commission. 2023. Delegated Regulation (EU) 2023/1184 of 10 February 2023 supplementing Directive (EU) 2018/2001 of the European Parliament and of the Council by establishing a Union methodology setting out detailed rules for the production of renewable liquid and gaseous transport fuels of non-biological origin. [2023] OJ L 157/11.

savings criterion established in Delegated Regulation (EU) 2023/1185.[33] In simple terms, only the hydrogen produced via an electrolyser powered by a renewable electricity generation plant installed or expanded *specifically* to feed the electrolyser, and that saves at least 70% of GHG emissions compared to the fuels it replaces, can qualify as RFNBO.[34]

Whereas, the legal definition for low-carbon hydrogen is 'hydrogen the energy content of which is derived from non-renewable sources, which meets the greenhouse gas emission reduction threshold of 70% compared to the fossil fuel comparator for renewable fuels of non-biological origin set out in the methodology for assessing greenhouse gas emissions savings from renewable fuels of non-biological origin and from recycled carbon fuels, adopted pursuant to Article 29a(3) of Directive (EU) 2018/2001.'[35] These two definitions are relevant due to the nascent nature of the hydrogen market in Europe. Its flourishing will require massive state resources. For instance, RFNBOs are eligible to receive state support under the Renewable Energy Directive (RED III) and can account for the renewable energy targets of Member States, whereas low-carbon hydrogen cannot.

This chapter focuses on this specific way of producing renewable hydrogen because the EU is prioritising its development.[36] It is important, however, to mention that this is not the only way of producing green hydrogen.[37] More importantly, it should be highlighted that other countries may legally define what constitutes 'green' or 'renewable' hydrogen differently.

[33] European Commission. 2023. Delegated Regulation (EU) 2023/1185 of 10 February 2023 supplementing Directive (EU) 2018/2001 of the European Parliament and of the Council by establishing a minimum threshold for greenhouse gas emissions savings of recycled carbon fuels and by specifying a methodology for assessing greenhouse gas emissions savings from renewable liquid and gaseous transport fuels of non-biological origin and from recycled carbon fuels. [2023] OJ L 157/20.

[34] For a further explanation see Talus, K.; Pinto, J. and Gallegos, F. 2024. Realism at the end of the rainbow? An argument towards diversifying hydrogen in EU regulation. *Journal of World Energy Law and Business, 17*(4): 217–233. https://doi.org/10.1093/jwelb/jwae007

[35] European Parliament, Directive (EU) 2024/1788 on common rules for the internal markets for renewable gas, natural gas and hydrogen, amending Directive (EU) 2023/1791 and repealing Directive 2009/73/EC, OJ L, 2024/1788, 15.7.2024 (RGH Directive).

[36] EU Green Deal; EU Hydrogen Strategy; EU Energy System Integration Strategy; RE-PowerEU Plan.

[37] Other ways include pyrolysis and biogas reforming. For a description of other ways of producing renewable hydrogen see: Martin Morán, A. 2024. Diversifying Clean Hydrogen. Regulatory Barriers for a Circular Hydrogen Economy. *Journal for European Environmental & Planning* Law, *21*(3–4), 167–192. https://doi.org/10.1163/18760104-21030003

Nonetheless, if its production is targeted towards trading it in the EU, it must then comply with the production standards established by the EU for RFNBOs as explained above. Therefore, if Tunisia's aim is to export green hydrogen to the EU, then its production must comply with EU rules.

To meet the RFNBO requirements, countries trading with the EU will have to make significant investments in developing renewable energy infrastructure, which can be especially taxing if they do not already have an existing robust energy system that already caters to local demand. The requirements described suggest that many infrastructures need to be developed in Tunisia. First and foremost, electrolysers. Second, new renewable electricity power plants will need to be installed, or operating renewable electricity power plants will need to be expanded with additional generation capacity (to comply with the RFNBO requirement). Third, an expansion of the electricity grids (transmission and/or distribution lines) may be needed to transport the new electricity that will power the electrolysers. Finally, to ensure the transport of hydrogen from one shore of the Mediterranean to the other, two possible options are either to transport it in its pure form through a dedicated cross-border maritime pipeline or, alternatively, via an energy carrier (such as ammonia) through a combination of shipping, train, and truck routes. Whereas the RED III is targeted to incentivise the production of RFNBOs, the EU is also incentivising the development of large-scale cross-border infrastructure, which is considered strategic for the development of the EU's internal market of energy (in this case, hydrogen) under the Projects of Common Interest (PCIs) and Projects of Mutual Interest (PMIs) framework.[38] PCIs and PMIs are projects that, after an exhaustive selection procedure, become part of the List of PCI[39] and PMIs that the EU adopts every two years. Projects included on the list benefit from a streamlined permit-granting procedure and the possibility of accessing financial assistance from the

[38] European Parliament and the Council Regulation (EU) 2022/869 of 30 May 2022 on guidelines for trans-European energy infrastructure, amending Regulations (EC) No 715/2009, (EU) 2019/942 and (EU) 2019/943 and Directive 2009/73/EC and (EU) 2019/944, and repealing Regulation (EU) No 347/2013 OJ L 152/45 ('TEN-E Regulation').

[39] The current list was published in Commission Delegated Regulation (EU) 2024/1041 of 28 November 2023 amending Regulation (EU) 2022/869 of the European Parliament and of the Council as regards the Union list of projects of common interest and projects of mutual interest, OJ L 2024/1041, 8.4.2024. Nonetheless, the next list is expected by the end of November 2025.

EU,[40] through the Connecting Europe Facility (CEF). Infrastructure with a union impact, such as cross-border hydrogen pipelines, hydrogen storage facilities, facilities for converting hydrogen from liquid to gaseous form and vice versa, and even the installation of electrolysers (that provide flexibility to the electricity system), may be eligible to become PCIs, whereas cross-border pipelines connecting the EU with a third country may be eligible to become PMIs whenever they fulfil the eligibility criteria.

In the context of this chapter, PMIs are relevant because cross-border infrastructure projects that are co-developed by an EU entity and a third-country entity may receive some of the benefits mentioned – including financial assistance under the CEF[41] – if they demonstrate the following requirements: (i) the project is located in at least one EU Member State and in another country outside the EU, (ii) the project does not hinder the capacity of the third country to phase out fossil fuel generation for its domestic consumption and to sustainability, and (iii) the potential overall benefits do not outweigh its costs in the longer term.[42] The H2GG projects analysed in this chapter, which are relevant for the development of the NAS Corridor, have either been included in the EU's list of PCIs and PMIs[43] or are candidates for the following the EU's list to be adopted by the end of 2025.[44]

[40] Article 1.2 of the European Parliament and the Council Regulation (EU) 2022/869 of 30 May 2022 on guidelines for trans-European energy infrastructure, amending Regulations (EC) No 715/2009, (EU) 2019/942 and (EU) 2019/943 and Directive 2009/73/EC and (EU) 2019/944, and repealing Regulation (EU) No 347/2013 OJ L 152/45 ('TEN-E Regulation').

[41] Articles 18 and 19 of the TEN-E Regulation and Article 5(2) of Regulation (EU) 2021/1153 of the European Parliament and of the Council of 7 July 2021 establishing the Connecting Europe Facility and repealing Regulations (EU) No 1316/2013 and (EU) No 283/2014 (OJ L 249, 14.7.2021.

[42] Article 4.2 of TEN-E Regulation.

[43] Four projects related to the NAS Corridor are currently part of the Union's list of PCIs and PMI, numbered 10.1. These include hydrogen pipelines and related infrastructure for the development of the Italian (10.1.1), Austrian (10.1.2 and 10.1.3) and German (10.1.4) hydrogen systems. Annex to the Delegated Regulation (EU) 2024/1041 of 28 November 2023 amending Regulation (EU) 2022/869 of the European Parliament and of the Council as regards the Union list of projects of common interest and projects of mutual interest, OJ L 2024/1041, 8.4.2024, and the Technical information on Projects of Common Interest and Projects of Mutual Interest, accompanying the Commission Delegated Regulation. Available at: https://energy.ec.europa.eu/document/download/944b96b9-4efd-44a3-bbfe-45b752b0b55f_en (Accessed: 10 October 2025).

[44] The section of the project corresponding to the maritime pipeline to connect Tunisia with Italy was included as a candidate project for the next Union's List of PCIs and PMIs

Thus, if a project is initiated as PMI, there is a commitment that the project will not only be sustainable and result in net socio-economic benefits for the EU and the third country, but that it will also not hinder or prevent the third country from engaging in a local green energy transition.[45] The authors argue that this means the H2GG projects listed as PMIs are thus obligated to demonstrate they do not prioritise the EU's green energy transition at the expense of the local green energy transition in third countries. This does not appear to impose a positive obligation on parties involved in the PMI, including Member States, infrastructure operators and legal entities based in the EU, to ensure that mutual benefits are realised, as the regulation only requires that the potential benefits from the PMI outweigh long-term cost. However, it does put a negative obligation on those involved in the project to prevent harm and to not place an unreasonable burden on third countries, as this is explicitly stated in the second criterion for PMIs identified above.[46] This is further affirmed by the fact that project promoters are obligated under the law to include in their annual report to the national regulators how the PMI complies with the principle of 'do no significant harm' of environmental law.[47] Additionally, the authors argue that by claiming that the Global Gateway and H2GG projects are sustainable and mutually beneficial for both the EU and its partner countries, the EU may have a duty.[48] This chapter will investigate whether this claim remains credible in the context of H2GG projects in Tunisia and if the EU is obligated to take measures to ensure that the benefit is for both regions and does not simply prioritise the EU's green transition at the expense of a global green transition.

7.2.2 Green Colonialism as a Counter-narrative

The EU's Global Gateway strategy aims at investing in global infrastructure projects with the aim of, among other goals, speeding up sustainable development, creating jobs, and assisting the transition to a cleaner and circular

as a PMI under the name North Africa Corridor. See the recorded version where hydrogen project promoters of the HI East Priority Corridor presented their projects. Available at: https://webcast.ec.europa.eu/hydrogen-pcipmi-east-meeting-2025-04-14 (Accessed: 10 October 2025).

[45] Recital 20 & Article 4.2(a) of TEN-E Regulation.

[46] Art 4. 2 (a) of the TEN-E Regulation.

[47] Art 5 .4 (a) of the TEN-E Regulation.

[48] European Commission. 2021. Joint Communication, JOIN(2021) 30 final, The Global Gateway.

global economy.[49] The EU affirms that its investments in energy infrastructure, including the H2GG projects, are aimed at both diversifying the EU's clean energy supply while also supporting the greening of local value chains and enabling local just energy transitions in third countries.[50] This can be construed as a claim by the EU that the H2GG projects in third countries, such as Tunisia, are sustainable and mutually beneficial.[51] However, this claim remains highly contested, as scholars, communities and activists have argued that projects that push for the export of renewable energy from North Africa to the EU, including green hydrogen projects, are actually forms of neocolonialism referred to as 'green colonialism'.[52] Some scholars argue that renewable energy projects are utilising the same inequitable systems as those used to extract fossil fuels for the benefit of high-income economies,[53] and will be accompanied by socio-environmental costs being displaced onto host countries and communities.[54] In some cases, such green colonialism manifests in third countries such as Tunisia through 'green [land/resources] grabbing', whereby large swathes of land are acquired for building solar power plants or wind farms on the basis that these desert lands are unusable for agricultural purposes and therefore considered available and ideal for such green projects.[55]

[49] Ibid.

[50] Ibid.

[51] Ibid.

[52] Hamouchene, H and Sandwell, K (eds). 2023. Dismantling Green Colonialism: Energy and Climate Justice in the Arab Region(1st edn, Pluto Press). Available at: http://www.jstor.org/stable/jj.7583922.7 (Accessed: 18 June 2025); Dejonghe, M. and Van de Graaf, T. 2025. Green Colonialism or Green Transformation? The Equity Implications of Clean Hydrogen Trade. *Political Geography*, 120, https://doi.org/10.1016/j.polgeo.2025.103338; Claar, S. 2022. Green Colonialism in the European Green Deal: Continuities of Dependency and the Relationship of Forces between Europe and Africa. *Culture, Practice & Europeanization, 7*(2): 262–274. DOI:10.5771/2566-7742-2022-2-262

[53] This defineation follows the World Bank country classification by income level. Metreau, E *et al.* 2024. World Bank country classifications by income level for 2024–2025. World Bank blogs. Available at: https://blogs.worldbank.org/en/opendata/world-bank-country-classifications-by-income-level-for-2024-2025 (Accessed: 20 October 2025)

[54] Hamouchene, H. 2023. The Energy Transition in North Africa: Neocolonialism Again! In: Hamouchene H. & Sandwell, K. (Eds.), *Dismantling Green Colonialism:* Energy and Climate Justice in the Arab Region (1st ed., pp. 29–48). Pluto Press. Available at: https://doi.org/10.2307/jj.7583922.7

[55] Hamouchene, H. *et al.* 2022. Renewable energy in Tunisia: An unjust transition. In: Hamouchene H. & Sandwell, K. (Eds.), *Dismantling Green Colonialism: Energy and Cli-*

Scholars such as Hamouchene argue that projects prioritising renewable energy exports halt or derail the local green transition of host countries, rather than facilitating it, precisely because they prioritise European energy policies over the interests of the local population.[56] The question then arises of whether all H2GG projects are predisposed to prioritise the EU energy goals over local demand if they are initiated with the intention of being primarily export-orientated, or if they have the capacity to facilitate the local green transition. Tunisian and international NGOs have argued that such projects, if pursued without democratic accountability or transparent planning, risk deepening inequalities rather than delivering a just energy *transition*.[57] This may mean that if there are sufficient safeguards, these H2GG projects could potentially facilitate the local transition or at the very least not hinder it. As El Amine highlights, just transition narratives must account for local realities, not only international demands.[58]

7.2.3 Green Hydrogen Policy and Projects in Tunisia

The EU has included several projects in the field of renewable energy and electricity interconnection with Tunisia as part of the Global Gateway.[59] This section will focus on the flagship projects of the Global Gateway that are part of the SoutH2 Corrido, and key for the development of the NAS Corridor of the EU H2 Backbone, mainly the development of the offshore

mate Justice in the Arab Region (1st ed., pp. 29–48). Pluto Press. Available at: https://doi.org/10.2307/jj.7583922.15

[56] Hamouchene, H. 2023. The Energy Transition in North Africa: Neocolonialism Again! In: Hamouchene H. & Sandwell, K. (Eds.), *Dismantling Green Colonialism: Energy and Climate Justice in the Arab Region* (1st ed., pp. 29–48). Pluto Press. Available at: https://doi.org/10.2307/jj.7583922.7

[57] Centre Tricontinental (CETRI). 2024. Green Hydrogen: A New Mechanism of Green Colonialism? Available at: https://www.cetri.be/Green-hydrogen-A-new-mechanism-of; Hamouchene H., & Sandwell, K. (Eds.). 2023. *Dismantling green colonialism: Energy and climate justice in the Arab region* (1st ed.). Pluto Press. https://doi.org/10.2307/jj.7583922

[58] El Amine, Y. 2025. Just Transition Green Bridge: Tunisian Case Study. Arab Reform Initiative. Available at: https://www.arab-reform.net/publication/just-transition-green-bridge-tunisian-case-study/. (Accessed: 20 October 2025).

[59] The Global Gateway. Information on the roll out of the Global Gateway strategy, partnerships, projects and funding opportunities. Available at: https://international-partnerships.ec.europa.eu/policies/global-gateway_en (Accessed: 28 May 2025).

pipeline that will connect the south of Italy with Tunisia and the development of large-scale hydrogen production projects in Tunisia for exporting it into the EU (hereinafter referred to as the 'SoutH2 Corridor').[60]

These projects would see Tunisia entering one of the largest energy markets in the world, that of the EU, which aims to become the first carbon-neutral continent. It is estimated that the export revenue from hydrogen will result in Tunisia gaining a positive trade balance of EUR 2.1 billion in 2035 and EUR 8.6 billion in 2050.[61] Additionally, green hydrogen projects are expected to create employment opportunities for the local population and increase foreign investment in Tunisia.[62] Another advantage of involving foreign partners from the EU is the potential transfer of technology and expertise in implementing clean and renewable energy.[63] In the absence of locally produced technology to implement projects on such a large scale, this can become a mutually beneficial position for both the EU and Tunisia.

However, despite these apparent benefits, the SoutH2 Corridor has been met with resistance from local civil society organisations, which warn that it risks replicating extractive development models (i.e. green colonialism), as it prioritises the EU's energy policy goals over the needs and rights of local communities.[64] Tunisia is also currently experiencing significant strains on its energy sector and has a highly volatile domestic energy market. It remains heavily reliant on non-renewable energy imports from

[60] North Africa Hydrogen Backbone. Sea Corridor. Available at: https://www.sea-corridor.com/en/h2-initiatives; https://www.south2corridor.net/hydrogen-production-te-h2-and-verbund-have-signed-a-mou-with-the-republic-of-tunisia/ Accessed: 13 August 2025).

[61] GFA & ALCOR. National Strategy for the Development of Green Hydrogen and its Derivatives in Tunisia. July 2024. Available at: https://www.energiemines.gov.tn/fileadmin/docs-u1/Re%CC%81sume%CC%81_strate%CC%81gie_nationale_MIME_Anglais.pdf (Accessed: 20 October 2025).

[62] GFA & ALCO. National Strategy for the Development of Green Hydrogen and its Derivatives in Tunisia. July 2024. Available at: https://www.energiemines.gov.tn/fileadmin/docs-u1/Re%CC%81sume%CC%81_strate%CC%81gie_nationale_MIME_Anglais.pdf (Accessed: 20 October 2025).

[63] Deutsche Gesellschaft für Internationale Zusammenarbeit (GIZ). 2022. Establishing a Bavarian-Tunisian Technology and Innovation Hub for Green Hydrogen. Available at: https://www.giz.de/en/downloads/giz2022-en-bavaria-tunisia.pdf (Accessed: 20 October 2025).

[64] Corporate Europe Observatory. 2025. We say no to the South H2 Corridor: A statement from 87 CSOs. Available at: https://corporateeurope.org/en/corporateeurope.org/en/2025/03/CSOs-reject-South-H2-corridor (Accessed: 20 October 2025).

Algeria, which account for over 50% of national demand.[65] Thus, there is a need to shift towards renewable energy production, and the intensified cooperation between the EU and North African countries on climate and energy policy, particularly under the frameworks of the Global Gateway, has created a greater momentum for hydrogen development and renewable energy production.[66]

For instance, to meet foreign energy demands, such as that of the EU, Tunisia has made plans to leverage its solar and wind resources, aiming to reach 35% of renewable electricity production by 2030 – against the 3% achieved in 2024.[67] Tunisia's National Hydrogen Strategy, published in 2024, set an ambitious production target of 8.3 million tonnes of green hydrogen annually by 2050. Of which, 2.3 million tonnes would be allocated for domestic use, and up to 6 million tonnes would be exported to the EU starting in 2030 via the SoutH2 Corridor.[68] It remains unclear if this increased renewable energy production would be used to first meet local demand or if supplying the EU with green hydrogen would be the priority. In its case study of the just transition discourse in Tunisia, the think tank *Arab Reform Initiative* showed that Tunisia's green hydrogen ambitions are driven largely by external demand rather than addressing domestic energy needs.[69] This ties in with the green colonialism argument, where the needs of the perceived 'metropole' (the EU) are prioritised

[65] Energy imports resulting in over 53.4% of Tunisia trade deficit in the first eight months of 2023, a figure which has doubled since 2017.World Bank Group. 2024.A virtuous cycle of green, affordable, and financially viable energy production. Available at: https://www.worldbank.org/en/programs/teri/brief/green-affordable-and-viable-energy-production-tunisia#:~:text=The%20Government%20of%20Tunisia%20(GoT,2030%2C%20against%203%25%20currently (Accessed: 20 October 2025).

[66] European Think Tanks Group (ETTG). A new chapter in the EU–North Africa Partnership on Climate and Energy. 2024. Available at: https://ettg.eu/eu-north-africa-partnership-climate-energy/ (Accessed: 20 October 2025).

[67] World Bank Group. 2024. A virtuous cycle of green, affordable, and financially viable energy production. Available at: https://www.worldbank.org/en/programs/teri/brief/green-affordable-and-viable-energy-production-tunisia#:~:text=The%20Government%20of%20Tunisia%20(GoT,2030%2C%20against%203%25%20currently (Accessed: 20 October 2025).

[68] Martin, P. 2024. Tunisia's new national green hydrogen strategy sets out plans for first H2 exports to Europe by 2030. Hydrogen Insight. Available at: https://www.hydrogen-insight.com/policy/tunisias-new-national-green-hydrogen-strategy-sets-out-plans-for-first-h2-exports-to-europe-by-2030/2-1-1651045 (Accessed: 20 October 2025).

[69] El Amine, Y. 2025. Just Transition Green Bridge: Tunisian Case Study. Arab Reform Initiative. Available at: https://www.arab-reform.net/publication/just-transition-green-bridge-tunisian-case-study/ (Accessed: 20 October 2025).

over those of the perceived 'periphery' (Tunisia). Tunisia, as a sovereign state, has the freedom to determine its energy policy. As mentioned above, however, it tends to prioritise the extension of its energy infrastructure to mainly supply the EU, an external market, rather than its domestic market.

Similarly, the burdens of investing and constructing infrastructure for this project do not appear to be equitably distributed. The production of green hydrogen would require significant investment to ensure that the necessary infrastructure is built in Tunisia, such as the vast installations of solar and wind electricity production plants, desalination plants, new pipelines for transporting hydrogen, new transmission grids, and port infrastructure development. The Tunisian Ministry of Energy and Mines and international actors frame these investments as a way to turn Tunisia into a 'green growth engine'.[70] However, for the transport and distribution of green hydrogen throughout the EU, the SoutH$_2$ Corridor will use a significant amount of repurposed pipeline (>65%) rather than completely new infrastructure.[71] This means that, unlike their Tunisian counterparts, the EU partners will be less burdened with investing in new infrastructure and can minimise the environmental harm that would typically arise from constructing new infrastructure. In contrast, Tunisia will have to invest in new infrastructure, which will be both costly and have a greater environmental impact. This further indicates that the burdens of green hydrogen supply chains are not equally borne by all parties.

Furthermore, given that Tunisia requires foreign investment for its green hydrogen infrastructure, it is unlikely that the profits from green hydrogen will remain within Tunisia. This makes such projects more likely to be exploitative of, rather than beneficial for, the energy-producing country. It is also unlikely that Tunisia will benefit heavily from the knowledge-sharing initiatives, as they remain small and are far from addressing more systemic barriers to innovation in Tunisia's higher education and research sectors.[72] Meaning that entering into this project will

[70] World Bank Group. 2024. A virtuous cycle of green, affordable, and financially viable energy production. Available at: https://www.worldbank.org/en/programs/teri/brief/green-affordable-and-viable-energy-production-tunisia#:~:text=The%20Government%20of%20Tunisia%20(GoT,2030%2C%20against%203%25%20currently (Accessed: 20 October 2025).

[71] SoutH2Corridor project website: Available at: https://www.south2corridor.net/ (Accessed: 18 October 2025).

[72] Allouche, B. and Abdeljalil, A. 2020. L'enseignement supérieur en Tunisie: A-t-on sacrifié la qualité face aux pressions quantitatives? In: Revue marocaine de la pensée contempo-

not necessarily result in Tunisia receiving assistance to build the necessary technical and professional capacity needed for its own local projects.

Presently, it is difficult to verify how this project will benefit the local population due to a lack of transparency. Despite the scale of the project, no clear estimates exist regarding direct or indirect local job creation from this project. Investment figures for the corridor also remain undisclosed, along with the final investment decisions, consultations with communities and environmental impact assessments, which remain publicly undisclosed.[73] These factors make it difficult to confirm whether the SoutH2 Corridor project has adequate safeguards against exploitative practices and if there are mechanisms to ensure that the benefits are equitably shared with Tunisia and its local population.

7.3 Energy Justice as a Framework for Assessing H2GG Initiatives

7.3.1 *Why Procedural Justice Matters*

To understand how such H2GG projects can avoid perpetuating green colonialism and instead facilitate the green transition in third countries, or at the very least be mutually beneficial, this section analyses the issue using the lens of energy justice. This lens will provide greater insight into which elements in a 3GH project are likely to ensure that the project is mutually beneficial and thus must be fostered, and which safeguards must be adopted to prevent exploitation and environmental harm.

Energy justice, as a normative concept, aims to ensure that the benefits and burdens of energy systems are distributed equitably across the different stages of the energy cycle and that the decisions made in the

raine, 2020, n° 5. Available at: https://archive-ouverte.unige.ch/unige:171798 (Accessed: 20 October 2025).

[73] Corporate Europe Observatory. 2025. We say no to the South H2 Corridor: A statement from 87 CSOs. Available at: https://corporateeurope.org/en/corporateeurope.org/en/2025/03/CSOs-reject-South-H2-corridor (Accessed: 20 October 2025); Ammar, E., & Ammar, S. 2024. Green hydrogen in Tunisia: A new mechanism of plunder and exploitation. Transnational Institute. Available at: https://www.tni.org/en/article/green-hydrogen-in-tunisia;El Amine, Y. Just Transition Green Bridge: Tunisian Case Study. Arab Reform Initiative. Available at: https://www.arab-reform.net/publication/just-transition-green-bridge-tunisian-case-study/ (Accessed: 20 October 2025).

global energy system are just, equitable, and representative.[74] There are different dimensions that form the concept of energy justice, namely distributive, procedural, recognition, restorative, and cosmopolitan.[75] The former two aspects have been extensively discussed in the literature, which is why the authors will focus on defining these two aspects.[76] Distributive justice is concerned with, as the name suggests, the equitable distribution of material outcomes from the energy system.[77] Whereas procedural justice is concerned with the processes involved in revealing, identifying and remediating energy injustices, including how decisions that impact society are made and who is involved in or can influence such decision-making.[78]

From an energy justice perspective, the issue of green colonialism and green hydrogen projects is one of injustice, where the parties and communities involved in the production phase of the energy lifecycle are disproportionately burdened with the costs of the energy systems, whereas the benefits of that system are enjoyed primarily by the parties involved in the distribution and consumption phases of the lifecycle. Furthermore, these injustices are worsened by decisions regarding the energy system being made by those outside the production phase, that

[74] Heffron, R. J. and Talus, K. 2016. The Evolution of Energy Law and Energy Jurisprudence: Insights for Energy Analysts and Researchers. *Energy Research & Social Science, 19*: 1–10. https://doi.org/10.1016/j.erss.2016.05.004; Jenkins, K. E. H. *et al.* 2021. The Methodologies, Geographies, and Technologies of Energy Justice: A Systematic and Comprehensive Review. *Environmental Research Letters*, 16 (4). DOI 10.1088/1748-9326/abd78c; Sovacool, B. K. and Dworkin, M. H. 2015. Energy Justice: Conceptual Insights and Practical Applications. *Applied Energy, 142*: 435–444. https://doi.org/10.1016/j.apenergy.2015.01.002

[75] Heffron, R. J. 2022. Applying Energy Justice into the Energy Transition. *Renewable and Sustainable Energy Reviews, 156*. https://doi.org/10.1016/j.rser.2021.111936

[76] Jenkins, K. E. H. *et al.* 2021. The Methodologies, Geographies, and Technologies of Energy Justice: A Systematic and Comprehensive Review. *Environmental Research Letters, 16*(4). DOI 10.1088/1748-9326/abd78c; Qian, Y. *et al.* 2023. A survey on energy justice: a critical review of the literature, *Economic Research-Ekonomska Istraživanja, 36* (3), 2155860. https://doi.org/10.1080/1331677X.2022.2155860

[77] Sovacool, B. K. and Dworkin, M. H. 2015. Energy Justice: Conceptual Insights and Practical Applications. *Applied Energy, 142*: 435–444. https://doi.org/10.1016/j.apenergy.2015.01.002

[78] Jenkins, K. E. H. *et al.* 2021. The Methodologies, Geographies, and Technologies of Energy Justice: A Systematic and Comprehensive Review. *Environmental Research Letters*, 16 (4). DOI 10.1088/1748-9326/abd78c; Sovacool, B. K. and Dworkin, M. H. 2015. Energy Justice: Conceptual Insights and Practical Applications. *Applied Energy, 142*: 435–444. https://doi.org/10.1016/j.apenergy.2015.01.002

is, EU entities and corporations. This overlap between the two concepts of green colonialism and energy justice was also observed and identified by Dejonghe and Van de Graaf.[79] They showed that domination and appropriation, which are key markers of green colonialism, align with the energy justice understanding of procedural and distributive injustice, respectively.[80] Scholars have noted that procedural injustices, such as asymmetrical decision-making and hegemonic power structures, are markers of green colonialism.[81]

In the previous section it was shown that Tunisia is likely to disproportionately bear the cost of these H2GG projects (both financially and environmentally) compared to the EU, and that the benefits of these projects, especially the access to clean energy, are largely enjoyed by the EU as opposed to Tunisia. Thus, we can infer from this that the H2GG projects, in addition to being engaged in green colonialism, are likely to also be energy unjust. We argue that this is likely the case due in part to the procedural injustice arising from the green hydrogen projects. This is because procedural injustice is one of the key elements of green colonialism (domination) and because procedural justice is instrumental in ensuring distributional justice. Without procedural justice, which acts as a safeguard, distributional injustices cannot be easily identified and addressed. For instance, procedural justice would require experts and communities who are most likely to shoulder the burdens of these energy systems to be involved in the decision-making processes. Failing to do so reduces the likelihood that injustices stemming from these projects will be identified and considered when decisions are made. Similarly, without redress mechanisms and complaint procedures that are required for procedural justice, any distributional injustices suffered by the local populace cannot be prevented or remedied.

[79] Dejonghe, M. and Van de Graaf, T. 2025.Green Colonialism or Green Transformation? The Equity Implications of Clean Hydrogen Trade. *Political Geography*, *120*. https://doi.org/10.1016/j.polgeo.2025.103338

[80] Ibid.

[81] Dejonghe, M. and Van de Graaf, T. 2025.Green Colonialism or Green Transformation? The Equity Implications of Clean Hydrogen Trade. *Political Geography*, *120*. https://doi.org/10.1016/j.polgeo.2025.103338; Hamouchene, H. 2023. The Energy Transition in North Africa: Neocolonialism Again! In: Hamouchene H. & Sandwell, K. (Eds.) Dismantling Green Colonialism: Energy and Climate Justice in the Arab Region (1st ed., pp. 29–48). Pluto Press. Available at: https://doi.org/10.2307/jj.7583922.7

7.3.2 Procedural Justice Safeguards and Mechanisms

Based on energy justice literature on procedural justice, the authors have identified five key procedural justice dimensions that can assist in ensuring that energy projects are environmentally sustainable and are not exploitative of the local communities that host these energy systems, especially those communities that are located at the production stage of these energy systems.[82] The five key dimensions are public participation and civil society; complaint and redress mechanisms; due diligence and accountability; access to information; and protection of vulnerable groups and equitable protection.

To illustrate how these justice dimensions will manifest in practice, the authors have identified several safeguards and mechanisms that can be incorporated into the operations of renewable energy projects, including the H2GG projects, as detailed in Table 7.1. These safeguards and mechanisms have been inspired by elements in the UN Framework Principles on Human Rights and the Environment.[83] This UN Framework in particular is important and relevant, as there are notable environmental and sustainability concerns surrounding these green energy projects due to the projects being in an ecosystem that already faces socio-environmental challenges such as water scarcity and increased desertification, and because there is a need for a local green transition.[84]

[82] Sovacool, B. K. and Dworkin, M. H. 2015. Energy Justice: Conceptual Insights and Practical Applications. *Applied Energy, 142:* 435–444. https://doi.org/10.1016/j.apenergy.2015.01.002; Heffron, R. J. 2022. Applying Energy Justice into the Energy Transition. *Renewable and Sustainable Energy Reviews,* 156. https://doi.org/10.1016/j.rser.2021.111936; Heffron, R.J. & de Fontenelle, L. (eds.) 2024. *The Power of Energy Justice & the Social Contract. Palgrave Macmillan.* https://doi.org/10.1007/978-3-031-46282-5; Forns Gómez, A., & Rehage, R. 2024. Is Public Participation in Hydrogen Matters (Un)just? Evaluating the Ex-/inclusion of the Vulnerable and Marginalised Individuals in the Hydrogen Decision-Chains of the EU, Spain and Germany. *Journal for European Environmental & Planning Law, 21*(3–4), 318–342. https://doi.org/10.1163/18760104-21030006; Sovacool, B. K. *et al.* 2019. Decarbonization and its discontents: a critical energy justice perspective on four low-carbon transitions. *Climatic Change, 155*:581–619. https://doi.org/10.1007/s10584-019-02521-7

[83] United Nations Human Rights Council. 2018. UN framework principles on human rights and the environment. A/HRC/37/59.

[84] Ecorys. 2013. Final Report - Trade Sustainability Impact Assessment in support of negotiations of a DCFTA between the EU and Tunisia. (Rotterdam). pp. 177–186. Available at: https://doi.org/10.13140/RG.2.1.1886.0400 (Accessed: 20 Ocotober 2025).

Table 7.1 Procedural Justice Safeguards and Mechanisms in Renewable Energy Projects.

Procedural Justice	*Safeguards and mechanisms.*
Public participation & civil society	• Ensuring that the local population is consulted or involved in decisions that can have significant effects on the environment and the human rights of the local population. • Requiring consultations with civil society, scientists and environmental and social justice activists for project decisions that can negatively affect the local population and the environment.
Complaint & redress mechanisms	• Creating mechanisms and procedures for the local population to lodge complaints when the operations of these projects affect their human rights or cause harm to the environment. • These mechanisms and procedures should be easily accessible and understandable to the local population. • There should be clear procedures for timely and adequate redress for complainants. • There should be no legal provisions that unreasonably restrict locals from taking legal action against the project operators and partners for harms to the environment or violations of human rights in local courts.
Due Diligence & Accountability	• Due diligence measures should be included, such as impact assessment, to prevent environmental harm and the infringement of human rights as a result of these projects. • The impact assessment should be reviewed by an independent body. • Requiring monitoring obligations to ensure that the proposed mitigation measures are implemented and to assess the effectiveness of these protective measures.
Access to information measures/rights	• Information on how human rights are affected by these projects, for example, from the impact assessments, should be made available to the public. • Obligation on project operators to collect comprehensive data on the environmental impact of these projects throughout their lifetime. • Obligations on project operators to ensure that these data are publicly available and easily accessible. • Obligations on project operators to ensure data access is affordable, effective, and timely.
Protection of vulnerable groups & equitable protection	• Requiring an assessment on how minority and vulnerable groups, especially those historically prejudiced and those that rely on ecosystems, are impacted by these projects. • Consulting experts to design safeguards to protect minority and vulnerable groups from suffering disproportionately due to the harms to the ecosystem.

Source (2025): Jimenez-Casanova, P.; Wickramanayake, S.A; Chabane, S.; Obrenovic, S. and Shah, S.A.A., [based on the UN Framework Principles on Human Rights and the Environment (A/HRC/37/59)].

This chapter argues that if procedural justice is ensured, especially in the initial stages of the project, the exploitative nature of these H2GG projects can be mitigated and minimised, and this can increase the likelihood of distributional justice, that is, increase the likelihood that the projects are in fact mutually beneficial and sustainable. Furthermore, the authors argue that if the safeguards and mechanisms detailed in Table 7.1 are shown to be present in the operations of these H2GG projects, it is likely that both procedural and distributional injustices will be mitigated. While all the components need not be present to ensure that the projects are not exploitative or extractive, the more elements there are, the greater the likelihood that injustices are eliminated. Adopting such safeguards and mechanisms can also show that if and when injustices are uncovered, parties to these projects have taken reasonable steps to mitigate such injustices. This may in turn enable these projects to potentially avoid legal liabilities and political backlash.

7.4 Why the EU Must Prevent Exploitation in Tunisia

7.4.1 Tunisia's Authoritarian Regression and Limits of Domestic Protection

As discussed in the previous sections, the H2GG projects, such as the SoutH2 Corridor, are already predisposed in their initial planning and development phases to be inequitable in how the burdens and benefits are distributed between Tunisia and the EU; that is, they potentially create distributional injustices. Ideally, the Tunisian government should negotiate for terms in the 3GH project agreements and trade agreements with the EU that would ensure that the needs and rights of the local population are met and protected, respectively, including terms to include procedural safeguards and mechanisms as detailed in Table 7.1. However, given both its historical colonial relationship (Tunisia's independence is less than 70 years old) and the current political realities, it is unlikely that communities can expect their government to ensure that there are adequate safeguards against energy injustices and the infringement of human rights. This section will examine the current political realities of Tunisia and explain why and to what extent this prevents Tunisians from being adequately protected against energy injustice and potential green colonialist projects.

In the aftermath of the Arab Spring and the overthrow of the Ben Ali regime, Tunisia made significant democratic gains, including by enshrining participatory governance practices in the new constitution.[85] These democratic gains, however, have been severely undermined since July 2021, when President Kais Saied suspended Parliament, dismissed the Prime Minister, and consolidated executive authority under his sole control.[86] Currently, Tunisia is classified as an electoral autocracy by the Varieties of Democracy Institute (V-DEM), meaning that Tunisia is 'substantially below' the 'starting levels' of democracies.[87] Tunisia is now facing a deepening crisis of democratic governance, given the significant erosion of its liberties, electoral processes and institutions, including judiciary and electoral authorities, under an emergency state.[88]

As explained in Section 3, procedural justice mechanisms can ensure distributional justice; that is, they can ensure a fair and equitable distribution of the benefits and burdens from the H2GG projects. However, these mechanisms usually require strong democratic governance for their enforcement. This mirrors the findings of Saadaoui *et al.*, who found that both democratic governance and institutional stability have a significant positive impact on the diffusion of renewable energy in Tunisia, whereas political instability and erosion of democratic norms hinder eco-friendly energy consumption.[89]

[85] OECD. 2016. Open Government in Tunisia, OECD Public Governance Reviews, OECD, pp. 82–118.

Publishing, Paris. http://dx.doi.org/10.1787/9789264227118-en (Accessed: 20 October 2025).

[86] Human Rights Watch. 2021. Tunisia: President's Seizure of Powers Threatens Rights: A dangerous move capitalizes on popular frustration. Available at: https://www.hrw.org/news/2021/07/27/tunisia-presidents-seizure-powers-threatens-rights (Accessed: 20 October 2025).

[87] V-DEM Institute. 2025. Democracy Report 2025. 25 years of Autocratization – Democracy Trumped?, p. 33. Available at: https://www.v-dem.net/documents/61/v-dem-dr__2025_lowres_v2.pdf (Accessed: 20 October 2025); Freedom House. 2024. Tunisia. Available at: https://freedomhouse.org/country/tunisia (Accessed: 20 October 2025).

[88] International Commission of Jurists. 2024. Tunisia: Unfair Presidential Election Undermines Human Rights and Deepens Political Crisis. Available at: https://www.icj.org/tunisia-unfair-presidential-election-undermines-human-rights-and-deepens-political-crisis/ (Accessed: 27 July 2025); Amara, T. 2024. Tunisia assembly votes to strip court of electoral authority, days before presidential vote. *Reuters*. Available at: https://www.reuters.com/world/africa/tunisia-assembly-votes-electoral-bill-nine-days-before-poll-opposition-calls-2024-09-27/ (Accessed 27 July 2025).

[89] Saadaoui, H. *et al.* 2024. The transition to renewable energies in Tunisia: The asymmetric impacts of technological innovation, government stability, and democracy. *Energy*, *293*. https://doi.org/10.1016/j.energy.2024.130686

Furthermore, strong democratic and independent institutions are usually needed for there to be adequate protection of human rights against the infringement of such rights. Without an independent legislature, there may be executive overreach, with governments making decisions and passing legislation that infringe and curtail human rights with minimal institutional pushback. Similarly, without an independent and strong judiciary, citizens are unlikely to get redress when their rights are infringed, especially if the state itself is the perpetrator of the infringement. Thus, the absence of democratic institutions in Tunisia makes it unlikely that state institutions will adequately protect citizens against green colonialism and its consequences on the local population. For instance, Tunisia's Ministry of Industry, Mines and Energy has already been accused of numerous violations of Law No. 12 of 2015, which governs private sector involvement in energy, including renewable energy projects.[90]

While there have been attempts to curtail civil society activities against hydrogen development,[91] Tunisia still has a vibrant civil society that is actively engaged in energy discussions. Local organisations (e.g., the Stop Pollution Movement, Voices for Just Climate Action) continue to advocate for a more socially just and sustainable model that prioritises citizens' needs and well-being.[92] Among them is the Working Group for Energy Democracy, which involves trade unions and other civil society actors in creating an alternative model for renewable energy and promoting 'energy democracy' over energy security.[93] Such public discussions will allow for energy injustices experienced by local communities to be identified and defined, which is necessary if redress is to be given.

[90] Hivos. 2025. Breakthrough for energy democracy in Tunisia. Available at: https://hivos.org/breakthrough-for-energy-democracy-in-tunisia/. (Accessed: 20 October 2025).

[91] Weipert-Fenner, I. 2025. Green Hydrogen Production in Tunisia: The Interplay of Old and New Lines of Conflict. Peace Research Institute Frankfurt. Available at: https://www.prif.org/fileadmin/Daten/Publikationen/Prif_Reports/2025/prif_2_25_barrierefrei.pdf (Accessed: 20 October 2025).

[92] Hivos. An insider's view on just climate action in Tunisia. Available at: https://hivos.org/story/an-insiders-view-on-just-climate-action-in-tunisia/ (Accessed: 20 October 2025); Ammar, I. 2022. Towards a just energy transition in Tunisia Transnational Institute. Available at:https://www.tni.org/files/2023-01/energy_democracy_tunesia_web.pdf (Accessed: 20 October 2025).

[93] Ammar, I. 2022. Towards a just energy transition in Tunisia Transnational Institute. Available at:https://www.tni.org/files/2023-01/energy_democracy_tunesia_web.pdf (Accessed: 20 October 2025).

Despite strong civil society engagement and public discourse, the chapter argues that the Tunisian legal system at present does not appear to have the necessary capacity to enforce procedural safeguards against harms stemming from the H2GG projects and ensure energy justice.

7.4.2 Does the EU Have a Duty to Prevent Harm?

Under international human rights law, all states have 'the *duty*' to protect against human rights abuses within their territory and/or jurisdiction by third parties committed by business enterprises'.[94] Applied to this example, it would mean that both EU Member States and Tunisia have a soft law obligation to prevent the exploitation and abuse of human rights that occur because of green hydrogen production in their territories. In addition, business enterprises have the '*responsibility* to respect human rights wherever they operate and whatever their size or industry', in other words, they should know 'their impacts, prevent and mitigate abuses, and address adverse impacts with which they are involved'.[95] However, as was established in the previous section, due to some Member States' prior colonial ties with Tunisia and Tunisia's weakened state of democratic institutions, it is unlikely that this responsibility will be faced, and safeguards will be adopted and enforced in Tunisia against possible exploitation and green colonialism that stem from green hydrogen projects. In this section, a case will be made for why the EU itself should be obligated to take measures to prevent its H2GG projects from exploiting Tunisia and ensure that there are safeguards against such exploitation.

On the level of EU member states there may be an obligation to regulate European enterprises involved in the H2GG projects. While the harms that result from these green hydrogen projects may not be within the EU territories, most of the parties involved are EU-based enterprises. In fact, several of the EU energy operators involved in the SoutH2 Corridor are (partially) owned and controlled by (local and national) public authorities and government agencies of member states.[96] According to

[94] United Nations, Guiding Principles on Business and Human Rights: Implementing the United Nations 'Protect, Respect and Remedy' Framework, Office of the High Commissioner for Human Rights, [ST/]HR/PUB/11/4, New York; Geneva, 2011, 2. Available at: https://digitallibrary.un.org/record/720245?ln=en (Accessed: 20 October 2025).

[95] Ibid.

[96] Italian operator Snam is controlled and owned by the Italian government indirectly through a holding company. Available at: https://www.snam.it/en/investor-relations/in-

the UN Guiding Principles on Business and Human Rights, states have an obligation under international human rights law to take steps to 'protect against human rights abuses by business enterprises that are owned or controlled by the State, or that receive substantial support and services from State agencies'. In light of these, EU member states have a soft law obligation to address the functioning of EU entities' operations to prevent them from engaging in human rights abuses.

The EU itself may also have similar soft law obligations. As these green hydrogen projects are also eligible for EU financing from the CEF,[97] EU institutions themselves are involved or at the very least facilitate the H2GG projects. If these projects are in fact exploitative and are forms of green colonialism, then the EU would be complicit in this extension of colonialism in the present, both financially and institutionally, as it would be facilitating the establishment of such projects to fulfil its policy objectives. This creates a misalignment with the EU's soft law obligations to promote global sustainability and prevent exploitation. These obligations stem from the 2016 Paris Agreement and the EU's own policies and foundational treaties.[98]

For instance, the EU in its foundational treaty has undertaken the task of promoting the protection of human rights and advancing the principles of equality and solidarity in its relations with the international community, as well as fostering sustainable economic, social and environmental development of 'developing countries'.[99] Similarly, the EU has also the mandate to promote measures dealing with global environmental issues through its environmental policies.[100] Moreover, the EU's Global Gateway strategy is intended to align with both the UN Sustainable Development

vestire-in-snam/azionariato.html (Accessed: 20 October 2025); Gas connect Austria is controlled by the Republic of Austria through the energy company Verbund that has a majority stake in gas connect. Available at: https://www.gasconnect.at/en/about-us/about-us/our-owners (Accessed: 20 October 2025); Baynernet is completely owned and controlled by German government institutions and municipalities. Available at: https://www.bayernets.de/en/about-us/who-we-are/shareholders (Accessed: 20 October 2025).

[97] Article 1.2 TEN-E Regulation.

[98] Paris Agreement of 19th October 2016. OJ L 282/4; Consolidated Treaty on European Union of 26 October 2012 OJ C 326; European Commission. 2021. Joint Communication, JOIN(2021) 30 final, The Global Gateway.

[99] Article 2, 3.5 & 21 of the Consolidated Treaty on European Union of 26 October 2012 OJ C 326.

[100] Article 191 of the Consolidated Treaty on the functioning of the European Union of 26 October 2012 OJ C 326.

Goals and the 2016 Paris Agreement. Thus, it is argued that the EU has a soft law obligation to align projects initiated under this strategy with international treaties. While soft law is not binding and enforceable in the same manner as its counterpart, or hard law[101] environmental policy. It is therefore argued that parties in the EU, or even Member States, may be able to hold the EU legally accountable, in case of failure to take measures to prevent and counteract environmental damage that would result from EU funded projects.

Many of these soft law obligations have already been incorporated in part into hard law by the EU through its legislation on corporate social responsibility and disclosure requirements.[102] However, given that the legislative area is currently being revised and will likely be applicable once these green hydrogen projects are complete and operational, this chapter will not explore the manner in which these due diligence and corporate responsibility obligations are imposed on the H2GG project promoters. A more direct link to the H2GG projects is the hard law obligations set down in the TEN-E Regulation.

As identified in Section 2, the EU enables financial assistance and administrative assistance to hydrogen projects that are classified as PCIs or PMIs.[103] Some projects included in the SoutH2 Corridor are already listed as PCIs under the TEN-E Regulation. Furthermore, the North Africa Hydrogen Backbone promoted by SeaCorridor, which is also part of the SoutH2 Corridor, is currently a candidate project to be included as a PMI in the next Union's List of PCIs and PMIs.[104] Therefore, these projects may receive assistance on the condition that they do not hinder the capacity of third countries to essentially engage in a local green energy transition nor negatively affect (environmental) sustainability. If these PMIs

[101] Feguš, V. 2023. Chapter 4: EU soft law: Validity, normativity and 'bindingness' reviewed. In The Legal Effects of EU Soft Law. Cheltenham, UK: Edward Elgar Publishing.

[102] European Parliament and Council. 2024. Directive (EU) 2024/1760 on corporate sustainability due diligence and amending Directive (EU) 2019/1937 and Regulation (EU) 2023/2859, OJ L 2024/1760, .05.07.2024 [CSDDD]; European Parliament and Council. 2022. Directive (EU) 2022/2464 amending Regulation (EU) No 537/2014, Directive 2004/109/EC, Directive 2006/43/EC and Directive 2013/34/EU, as regards corporate sustainability reporting, OJ L 322, 16.12.2022 [CSRD].

[103] Article 1.2 of the TEN-E Regulation.

[104] The North Africa Hydrogen Backbone, promoted by SeaCorridor is currently a candidate project to be included as a PMI for the following Union list to be adopted in the last quarter of 2025. Available at: https://www.sea-corridor.com/en/h2-initiatives (Accessed:13 August 2025).

do breach these conditions, the EU can and should hold European entities involved in the project accountable and legally liable. As explained in Sections 3 and 4, if the H2GG projects are found to be highly extractivist and perpetuate economic dependency – engaging in green colonialism – it can be assumed that they are in breach of these conditions. Furthermore, EU regulation already imposes obligations on the projects designated as PCIs to take steps to ensure many of the procedural justice dimensions identified in Table 7.1, such as public participation and transparency. Though these obligations appear to apply primarily with regard to operations and communities in the EU, rather than in the third countries. Therefore, requiring that these procedural justice safeguards and mechanisms are also followed in third countries, especially when they are listed as PMIs, would simply be an extension of the already existing obligations imposed on EU corporations and infrastructure operators.

7.4.3 MOUs and Agreements as Safeguard Instruments

Having established that there is both a moral and ethical obligation for the EU to take some measures to mitigate and prevent green colonial practices stemming from the H2GG projects, this section will discuss how the EU can potentially fulfil these obligations through MoUs and trade agreements.

The legal basis and general framework for cooperation between Tunisia and the EU are governed by the Association Agreement, which entered into force in March 1998.[105] Despite Tunisia's worsening democratic landscape, the EU also signed an MoU with the country in July 2023, which includes cooperation for a green energy transition,[106] and in 2024 entered into an MoU on renewable energy with the EU.[107] Such cooperation has been decried by the European Parliament, which approved a resolution through an urgent written procedure, as it condemned the

[105] Euro-Mediterranean Agreement establishing an Association between the European Communities and their Member States, of the one part, and the Republic of Tunisia, of the other part, *Official Journal of the European Union*, L 97, 30 March 1998.

[106] European Commission. 2023. Memorandum of Understanding on a strategic and global partnership between the European Union and Tunisia. Available at: https://ec.europa.eu/commission/presscorner/detail/en/ip_23_3887 (Accessed: 23 May 2025).

[107] EU Neighbours South. 2024. Tunisia: PPP, ELMED and the ambitious MoU Toward Green Hydrogen and Renewable Energy. Available at: https://south.euneighbours.eu/news/tunisia-ppp-elmed-and-the-ambitious-mou-toward-green-hydrogen-and-renewable-energy/ (Accessed: 15 May 2025).

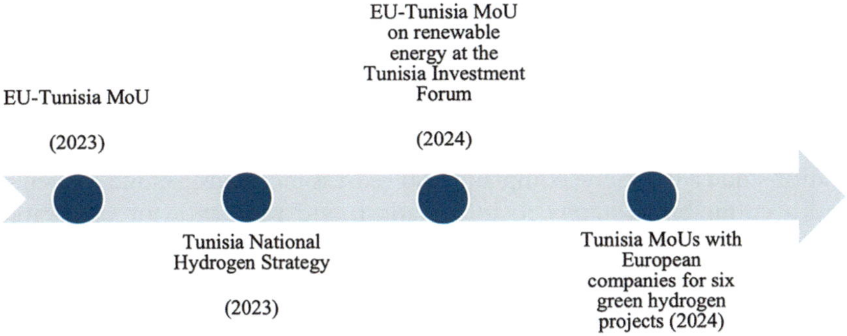

Fig. 7.1 Timeline of documents and policies for green hydrogen development in Tunisia. Source (2025): Created by Jimenez-Casanova, P.; Wickramanayake, S.A; Chabane, S.; Obrenovic, S. and Shah, S.A.A. Based on European Commission. 2023. Memorandum of Understanding on a strategic and global partnership between the European Union and Tunisia. Available at: https://ec.europa.eu/commission/presscorner/detail/en/ip_23_3887 (Accessed: 23 May 2025); Ammar, S. 2024. Resisting the new green colonialism. Transnational Institute. Available at: https://www.tni.org/en/article/resisting-the-new-green-colonialism (Accessed: 15 May 2025); EU Neighbours South. 2024. Tunisia: PPP, ELMED and the ambitious MoU Toward Green Hydrogen and Renewable Energy. Available at: https://south.euneighbours.eu/news/tunisia-ppp-elmed-and-the-ambitious-mou-toward-green-hydrogen-and-renewable-energy/ (Accessed:15 May 2025).

Commission's financial and diplomatic support of Tunisia.[108]Tunisia has also entered a number of additional MoUs with European companies for six specific green hydrogen projects in July 2024,[109] as illustrated in Fig. 7.1, all of these political agreements are key to paving the way for future projects and cooperation. However, the implementation of large projects will depend on the actual subscription of contracts and making final investment decisions.

[108] European Parliament resolution of 14 March 2024 on the adoption of the special measure in favour of Tunisia for 2023 (2024/2573(RSP)), P9_TA(2024)0175, 14 March 2024 https://www.europarl.europa.eu/doceo/document/TA-9-2024-0175_EN.html; Statewatch. 2024. Parliamentary lawyers: democratic oversight needed for EU-Tunisia migration agreement. Available at: https://www.statewatch.org/news/2024/march/parliamentary-lawyers-democratic-oversight-needed-for-eu-tunisia-migration-agreement/ (Accessed: 20 October 2025).

[109] Hydrogen Europe. 2024. Six hydrogen production deals concluded in Tunisia. Available at: https://hydrogeneurope.eu/six-hydrogen-production-deals-concluded-in-tunisia/ (Accessed: 12 May 2025).

One way the EU can take measures to prevent exploitation through the H2GG projects is to incorporate safeguard provisions that tackle the five key procedural justice dimensions in Table 7.1 into the agreements and the MoUs entered into by the EU and Tunisia, as well as those between Tunisia and private EU corporations. The five key tenets are public participation and civil society; complaint and redress mechanisms; due diligence and accountability; access to information; and protection of vulnerable groups and equitable protection. Even though incorporating safeguard provisions into agreements may be more difficult as the EU itself is not a party to those agreements. However, given that the PMI and PCIs receive financial and administrative support from the EU, it may be possible for the EU to mandate the incorporation of safeguards as part of the conditions for financing or issuing permits.

7.5 Conclusions

This chapter examined the potential energy justice issues that can arise from green hydrogen initiatives in Tunisia as part of the EU's decarbonisation strategy. It discussed how the counter-narrative of green colonialism appears as a valid concern raised by scholars and communities, particularly in relation to green hydrogen projects. Despite the EU's claims that the Global Gateway and H2GG projects are 'mutually beneficial', benefiting the economies of involved third countries and their local populations alongside responding to the EU's needs in matters of energy and climate targets, H2GG projects in Tunisia were shown to present risks of being exploitative. This risk was even more visible in the context of democratic backsliding and in contexts where socio-economic patterns of colonial relations persist in the present.

It was argued that by adopting an energy justice perspective in its projects, the EU can help ensure that H2GG projects are mutually beneficial and not exploitative. It has been shown that the adoption of procedural justice safeguards, in particular, can identify and prevent green colonialism and help facilitate, or at the very least not hinder, Tunisia's energy transition. The primary responsibility for ensuring that procedural safeguards are embedded in H2GG projects was shown to lie with the EU, while highlighting its necessity for Tunisia, given its deteriorating political climate. Finally, it ended with recommendations on which key elements of procedural justice should be incorporated into agreements between

Tunisia, the EU and European companies for the Global Gateway and H2GG projects. The analytical framework developed in this chapter could also be mobilised by other scholars to study different cases of green energy projects beyond Tunisia, particularly other North African countries such as Morocco and Algeria, given their shared historical ties with Europe and similar socio-environmental concerns, together with their engagement in the NAS H2 Corridor and the Global Gateway strategy.

PART V

Conclusion

CHAPTER 8

Applying a Critical Reflective Framework to Energy Justice: Piloting the 'What is the Problem Represented to be' Framework in an Academic Context

Sil Katrien Maslov-Lescrauwaet, Klaudia Szabelka, Rachel Walters, Tom Hambley, and Päivi Tikkakoski

Abstract This chapter applies Carol Bacchi's 'What's the Problem Represented to be?' (WPR) framework to examine how early career researchers (ECRs) conceptualise the 'problem' that energy justice aims to address. Through a collective analysis of chapters from the edited collection

S. Katrien Maslov-Lescrauwaet
United Nations University, Institute on Comparative Regional Integration Studies

K. Szabelka
Glasgow Caledonian University, Scotland, The United Kingdom
e-mail: klaudia.szabelka.gcu.ac.uk

R. Walters
Monash University, Melbourne, Australia

T. Hambley
Anglia Ruskin University, London

P. Tikkakoski
University of Helsinki, Finland

A. Phillips, L. Kaschny (eds.), *Global Perspectives on the Circular Economy in the Just Transition*, Just Transitions,
https://doi.org/10.1007/978-3-032-18342-2_8

New Engagement Strategies for Energy Justice, dominant representations, underlying assumptions, silences, and discursive effects within emerging energy justice scholarship are identified. The findings reveal that energy injustices are predominantly framed as technical or regulatory deficiencies, grounded in Western-centric, liberal, market-based assumptions. These representations risk reducing complex injustices to measurable processes, marginalising affected communities' voices, and reinforcing dominant paradigms. While analytical frameworks can support a more comprehensive analysis, their narrow application risks oversimplifying socio-political dynamics into manageable technical problems. It is argued that such narrow framings by scholars and, consequently, policymakers, reflect a prevailing dominant discourse about the causes, effects and required instruments of energy justice, and how energy transition governance should proceed. These risks leave systematic socio-economic injustices untouched. The chapter calls for moving beyond technocratic approaches towards recognising energy justice as a participatory, dynamic, and contested process that requires more attention to structural, epistemic, and political dimensions.

Keywords Energy justice • Problem representation • Early Career Researchers • Justice Narratives • Technocratisation • WPR framework

8.1 Introduction

Energy justice has emerged as a key concept in academic,[1] policy,[2] and applied[3] discourse in energy transitions, particularly on the societal transformations potentially enabled towards more sustainable energy system.[4] A major societal transformation that scholars advocate for in

[1] Jenkins, K. *et al.* 2016. Energy justice: A conceptual review. *Energy Res. Soc. Sci., 11*, 174–182. https://doi.org/10.1016/j.erss.2015.10.004; Bouzarovski, S. and Simcock, N. 2017. Spatializing energy justice. *Energy Policy, 107*, 640–648.

[2] United Nations. 2015. Paris Agreement. [ONLINE] Available at: https://unfccc.int/sites/default/files/english_paris_agreement.pdf (Accessed: 20 July 2025).

[3] Milčiuvienė, S. *et al.* 2019. The role of renewable energy prosumers in implementing energy justice theory. *Sustainability, 11*(19), 5286.

[4] Parkhill, K. *et al.* 2013. Transforming the UK energy system: public values, attitudes and acceptability-synthesis report; Stephens, J.C. 2019. Energy democracy: redistributing power to the people through renewable transformation. *Environment: Science and Policy for Sustainable Development, 61*(2), 4–13; Butler, C., Parkhill, K.A. and Luzecka, P. 2018. Rethinking energy demand governance: Exploring impact beyond 'energy'policy. *Energy Research & Social Science, 36*, 70–78.

sustainable energy transitions is addressing structural inequalities within socio-technical systems. In energy systems, there is inequality in energy consumption, energy access, and energy affordability,[5] with disparities existing on regional, national, and global scales.[6] Policy initiatives driving an energy transition will not address these inequalities without specific consideration and often themselves risk reinforcing unjust processes. Different cross-sections of society are 'unevenly and disproportionately' impacted by new energy technologies,[7] and there is often disjuncture between decarbonisation visions and the actions expected of individuals and the actions undertaken by those individuals.[8] This is particularly the case when policy initiatives fail to recognise community heterogeneity, which can reinforce injustices through social structures which misrepresent inequalities and social processes.[9]

For a transition to be considered just, explicit consideration will therefore need to be made of existing injustices, where new injustices emerge, which sections of society are excluded, and how those injustices can be addressed.[10] This will include not only the fair and equitable distribution of benefits and burdens but also ensuring inclusive and equitable representation in decision-making.[11] Energy justice is commonly applied with

[5] Khan, F. and Heinecker, P. 2018. Inequality and energy: Revisiting the relationship between disparity of income distribution and energy use from a complex systems perspective. *Energy Research & Social Science*, *42*, pp. 184–192; Volodzkiene, L. and Streimikiene, D. 2023. Energy inequality indicators: a comprehensive review for exploring ways to reduce inequality. *Energies*, *16*(16), 6075.

[6] McCauley, D. 2018. Global energy justice: tackling systems of inequality in energy production and consumption. *Energy Justice: Re-Balancing the Trilemma of Security, Poverty and Climate Change*, 1–26.

[7] Scott, M. and Powells, G. 2020. Towards a new social science research agenda for hydrogen transitions: Social practices, energy justice, and place attachment. *Energy Research & Social Science*, *61*, 101346.

[8] Crowther, A., Petrova, S. and Evans, J. 2023. Between vision and implementation: the exclusionary disjuncture of domestic heat decarbonisation in Greater Manchester. *Local Environment*, *28*(8), 1045–1061.

[9] Fraser, N. and Honneth, A. 2003. *Redistribution or recognition?: a political-philosophical exchange*. verso.

[10] Crowther, A., Petrova, S. and Evans, J. 2023. Between vision and implementation: the exclusionary disjuncture of domestic heat decarbonisation in Greater Manchester. *Local Environment*, *28*(8), 1045–1061; Jenkins, K. *et al.* 2016. Energy justice: A conceptual review. *Energy Research & Social Science*, 11, 174–182.

[11] Sovacool, B. K. and Dworkin, M. H. 2015. Energy justice: Conceptual insights and practical applications. *Applied energy*, *142*, 435–444.

evolving typologies. Where early work proposed various approaches, from two 'principles',[12] to eight 'aspects'[13] later expanded to ten 'principles',[14] the field has largely coalesced around dominant frameworks. The most established are the three 'tenets'[15] of procedural, distributional, and recognition justice, recently updated to include restorative and cosmopolitan dimensions to form five 'principles'.[16] However, this definitional evolution, diversity and complexity has increased the difficulty of applying energy justice as a concept to the policy sphere,[17] and reduced conceptual clarity. For example, interdisciplinary disagreements between scholars of energy justice and energy democracy[18] have resulted in both terms and their typological frames[19] being accused of becoming 'buzzwords',[20] and despite empirical research dominating energy justice literature, it is often insufficiently informed by theoretical research, with papers often referring to concepts and citing definitions but not theoretical frameworks with empirical findings.[21] As such, scholars have called for greater attention to be paid to critical reflection in energy justice scholarship to ensure that energy justice does not become a 'conceptual toolkit'[22] that fails to 'fully

[12] Pellegrini-Masini, G., Pirni, A. and Maran, S. 2020. Energy justice revisited: A critical review on the philosophical and political origins of equality. *Energy Research & Social Science, 59*, 101310.

[13] Sovacool, B. K. and Dworkin, M. H. 2015. Energy justice: Conceptual insights and practical applications. *Applied energy, 142*, 435–444.

[14] Sovacool, B. K. *et al.* 2017. New frontiers and conceptual frameworks for energy justice. *Energy policy*, 105, 677–691.

[15] Heffron, R. J. and McCauley, D. 2017. The concept of energy justice across the disciplines. *Energy policy, 105*, 658–667.

[16] Heffron, R. J. 2024. Energy justice–the triumvirate of tenets revisited and revised. *Journal of Energy & Natural Resources Law, 42*(2), 227–233.

[17] Ramazan, S. *et al.* 2017. *Energy justice – a social sciences and humanities cross-cutting theme report. Cambridge: SHAPE ENERGY*

[18] Droubi, S., Heffron, R. J. and McCauley, D. 2022. A critical review of energy democracy: A failure to deliver justice? *Energy Research & Social Science, 86*, 102444.

[19] Osička, J., Szulecki, K. and Jenkins, K. E. 2023. p14 Energy justice and energy democracy: Separated twins, rival concepts or just buzzwords?. *Energy Research & Social Science, 104*, 103266.

[20] Ibid.

[21] Ibid.

[22] van Uffelen, N. 2024. p1 Understanding energy conflicts: From epistemic disputes to competing conceptions of justice. *Energy Research & Social Science, 118*, 103809.

understand the core of energy justice conflicts',[23] and one that moves beyond making normative claims and towards practical integration.[24]

In this chapter, we propose Carol Bacchi's 'What Is The Problem Represented To Be' (WPR) framework[25] as a reflexive tool for scholars to critically reflect on their own work, the wider discourse, and the application of energy justice to policy and practice. In doing so, we explore whether the WPR framework could serve as a reflexive tool for the energy justice field to move beyond normative claims or serve as a checklist. To do this, we apply the framework to analyse all chapters from the edited volume *New Engagement Strategies for Energy Justice*.[26] The book represents a collection of interdisciplinary contributions to the energy justice field from ECRs and, as a curated collection, serves as a contained case study of some of the most recent thinking in the field. The collection therefore serves as an appropriate case study for piloting the WPR framework to explore whether it is suitable as a wider reflexive tool for the wider field.

Therefore, this chapter is guided by the following research questions:

1. To what extent does the WPR framework serve as a useful reflexive tool for energy justice scholars?
2. How do early career researchers (ECRs) represent the 'problem' that energy justice aims to address in their scholarship, and what assumptions, silences, and implications emerge from these representations?

8.2 Methodology

8.2.1 Research Aim and Methodological Framework

This study analyses whether Carol Bacchi's WPR framework could serve as a reflexive tool for energy justice scholars to review how they are representing energy justice in their work, how the concept is being represented

[23] van Uffelen, N. 2024. p1 Understanding energy conflicts: From epistemic disputes to competing conceptions of justice. *Energy Research & Social Science, 118*, 103809.

[24] Rainard, M., Büchs, M., Jenkins, K., & Middlemiss, L. (2025). Intersectionality in good faith: Beyond normative claims and towards practical integration in energy justice research. *Energy Research & Social Science, 119*, 103823.

[25] Bacchi, C. 2012. Introducing the 'What's the Problem Represented to be?'approach. *Engaging with Carol Bacchi: Strategic interventions and exchanges*, 21–24.

[26] Phillips, A. and Kaschny, L. (Eds.). *New Engagement Strategies for Energy Justice: Perspectives from the Next Generation. Springer Nature Switzerland.*

by their peers and the wider literature, and signpost the developmental direction of the field. In doing so, we adopt a critical analytical perspective to explore how ECRs conceptualise and represent the 'problem' that energy justice aims to address in their academic work.

Originally developed for policy analysis, the WPR approach examines how 'problems' are constructed through language and discourse.[27] In addition, it delves into the underlying assumptions that make particular problem representations comprehensible and explores the effects these portrayals produce.[28] While WPR has been applied to various policy domains, such as welfare reform[29] policy, its application to energy justice scholarship is underexplored. Yet, energy justice is particularly suited to WPR analysis because different stakeholders may frame 'the problem' very differently, revealing which framings become dominant and which alternatives are marginalised.

The choice of WPR as a method is deemed suitable because it allows for a critical analysis of commonly accepted assumptions within the academic discourse about energy justice, moving from the theoretical debates towards a more profound examination of how justice itself is defined as a subject of understanding and intervention. In addition, the WPR framework is grounded in post-structuralism, particularly Foucauldian discourse analysis, which examines how discourses function as power systems that shape subjects and knowledge instead of simply mirroring pre-existing fixed truths.[30] It sheds light on the idea that what we perceive as natural or inevitable is actually historically shaped and politically constructed.[31] This perspective enables a critical examination of the categories, assumptions, and narratives that are often taken for granted.[32] In the context of energy justice research, this theoretical framework can provide tools to question how justice is defined, who defines it, and what implications these definitions carry for research and practice. In this sense, the

[27] Bacchi, C. 2009. *Analysing Policy: What's the problem represented to be? Introduction. Pearson Australia.* ix–xxii

[28] Ibid.

[29] Ibid.

[30] Khan, T. H. and MacEachen, E. 2021. Foucauldian Discourse Analysis: Moving Beyond a Social Constructionist Analytic. *International Journal of Qualitative Methods, 20*, 16094069211018009. https://doi.org/10.1177/16094069211018009

[31] Ibid.

[32] Bacchi, C. and Goodwin, S. 2016. *Introduction. In Poststructural policy analysis: A guide to practice* (pp. 3–12). Springer.

framework aligns with the growing calls in energy justice research for reflexivity and action.[33,34] Lastly, the WPR's focus on power dynamics and knowledge production aligns with decolonial and feminist perspectives, which are increasingly being recognised.[35] Thus, though originating in Global North contexts, WPR is applied here with sensitivity to its origin. Its focus on discourse, power, and knowledge production aligns with post-colonial and decolonial methodologies, making it suitable for uncovering how Global North frameworks may marginalise or universalise energy justice narratives at the expense of alternative epistemologies.

8.2.2 Case Selection and Scope

The analysis focuses on the edited collection New Engagement Strategies for Energy Justice,[36] produced by ECRs examining energy justice across various technological, geographical, and policy landscapes and contexts. The selection of these cases serves different analytical goals: ECRs often occupy a distinct position in academic research, as they bring emergent perspectives while working within academic structures and publication expectations that may influence their conceptual framing. The chosen book includes nine chapters covering contexts ranging from nuclear energy governance to solar photovoltaic policies, and from corporate regulation to regional development initiatives. This variety of topics allows for an examination of how issues of energy justice are portrayed across different scales, technologies, and diverse locations, while pinpointing recurring discursive patterns that go beyond specific cases.

8.2.3 Analytical Framework: The WPR Approach

The WPR framework consists of six guiding questions, each of which probes a different dimension of problem representation (Table 8.1). These include identifying the problem, the way in which the problem is represented, uncovering underlying presuppositions or assumptions, tracing the

[33] Jenkins, K., McCauley, D., Heffron, R., Stephan, H., & Rehner, R. 2016. Energy justice: A conceptual review. *Energy Research & Social Science*, *11*, 174–182.

[34] Sovacool, B. K. and Dworkin, M. H. 2015. Energy justice: Conceptual insights and practical applications. *Applied Energy*, *142*, 435–444.

[35] Sultana, F. 2022. Critical climate justice. *The Geographical Journal*, *188*(1), 118–124.

[36] Phillips, A. and Kaschny, L. (Eds.), *New Engagement Strategies for Energy Justice: Perspectives from the Next Generation. Springer Nature Switzerland.*

Table 8.1 WPR Framework: Questions and Purposes.

WPR Question	*Purpose of the Question*
1. What's the problem represented to be?	To identify the implied problem representation in specific policies.
2. What presuppositions or assumptions underlie this representation of the 'problem'?	To identify and analyse the conceptual logics that underpin specific problem representations. The term 'conceptual logic' refers to meanings that must be in place for a particular problem representation to cohere or to make sense.
3. How did this representation of the 'problem' come about?	To highlight the conditions that allow a particular problem representation to take shape and to assume dominance.
4. What is left unproblematic in this problem representation?	To raise reflection and consideration about issues and perspectives silenced.
5. What effects are produced by this representation of the 'problem'?	To identify the effects of specific problem representations so that they can be critically assessed.
6. How/where has this representation of the 'problem' been produced, disseminated and defended?	To pay attention both to the means through which some problem representations become dominant and to the possibility of challenging problem representations that are judged to be harmful or limiting.

Source: Adapted from Bacchi (2009)[1]

[1] Bacchi, C. 2009. *Analysing Policy: What's the problem represented to be? Introduction. Pearson Australia.* ix-xxii

emergence and defence of the (problem) framings, the effects produced by certain representations, and examining what remains unproblematised or excluded.

Questions 1 and 2 focus on understanding the logic behind the ways in which problems are represented and interpreted. Question 3 uncovers the historical factors that give rise to specific framings and their prevalence. Question 4 explores silences and alternatives, whereas Questions 5 and 6 examine the effects of the identified representations and potentials for change.

8.2.4 Data Analysis Process

This analysis proceeded through two phases, each building upon the other to balance depth and breadth while maintaining analytical rigour. In the first phase, each chapter was analysed separately, through close critical reading guided by the WPR framework. This involved identifying explicit and implicit problem representations, mapping underlying assumptions that made these representations intelligible, and noting recurring themes, metaphors, and analytical strategies in each individual chapter. Particular attention was paid to how each chapter constructed its central concerns, how energy justice was defined, and which dimensions were emphasised (or overlooked). The part of the analysis was conducted in three parts. Initially each researcher created chapter summaries independently in a shared table, addressing each question of the WPR framework per separate chapter. These summaries were collectively reviewed to spot initial patterns. Next, researchers were tasked with addressing WPR questions related to the collection as a whole; their responses were compiled into a structured table and discussed to synthesise findings across the edited collection. Lastly, a final table was used to capture emerging insights from the results and reflections on the application of the WPR framework.

This was followed by group discussions to consolidate interpretations and fine-tune the analysis. For instance, it was noticed that chapters often used phrases like 'energy justice requires…' or 'the challenge is to….' Such language may portray energy justice as an external goal or solution that experts or policymakers should provide to communities, instead of viewing it as an ongoing and evolving process that can be shaped by community participation and self-determination. This linguistic framing positions affected communities in a passive position, rather than as active participants in shaping and creating justice. These patterns pointed implicitly towards a positioning of justice as something to be delivered, rather than co-created. The synthesis phase represented the application of the six WPR questions as macro-thematic lenses for examining the full collection. Here, each WPR question guided a complete re-engagement with the corpus. For example, during the mapping of underlying assumptions (Question 2), recurring conceptual frameworks that structured thinking across chapters were identified. Throughout both phases, analytical reflexivity was maintained through regular team discussions, peer review of coding decisions, and explicit attention to our own positionality as ECRs examining early career work. This reflexive approach acknowledges that the analysis is itself a form of problem representation, subject to the same critical scrutiny applied to the case material.

8.2.5 Results

The results presented in Table 8.2 are used to capture the emergent themes and discursive tendencies across the analysis. Across the chapters,

Table 8.2 WPR Analysis results of book chapters.

WPR Questiony	*Summary of Representation*	*Core Themes Identified*	*Example of Supporting Representation arising from Chapter Analysis*
Q1: What's the problem represented to be?	Energy justice is often represented as a technical or regulatory deficiency, with limited engagement with the political and structural dimensions embedded in these decisions.	• Disproportionate access (chs. 2, 3) • Efficiency prioritised over equity (chs. 3, 5) • EJ is treated as a tick-box exercise, rather than a transformative framework (ch. 6) • Focus on technical deployment, not structural causes (chs. 2, 7, and 8)	Chapter 5 examines a 2022 Polish proposal on Power-to-Gas (P2G), allowing system operators to convert energy from electrons to molecules. It frames the issue as regulating P2G to prevent monopolies and enable participation. By creating a new 'electronic conversion' category, the policy offers a regulatory workaround rather than treating the problem as classification.
Q2: What assumptions underlie this representation?	Justice is deliverable within existing structures; technocratic, economic, and liberal assumptions dominate.	• Status quo suffices: Justice framed within existing systems with limited interrogation of whether these systems themselves need to change (chs. 2, 3, 7) • Quantifiability: Justice as a measurable construct (chs. 2, 6). • Market logics: Tech and corporate actors led (chs. 5, 8).	In chapter 4 technological optimism assumes nuclear expansion can solve climate challenges if properly regulated. Procedural inclusion treats stakeholder engagement as a tool to build legitimacy. Justice is framed as measurable and universally applicable through regulatory reforms. International cooperation is assumed to mitigate risks, overlooking power asymmetries and local contexts. Recognitional justice relies on compensation and regulation to address historical harms.

		• Tech optimism: Decarbonisation overrides ethics (chs. 2 and 4). • Procedural > Recognitional justice (chs. 4 and 6). • Typology rigidity: Universal justice typologies applied in siloed, Global North terms (chs. 2, 6, and 7).	
Q3: How has this representation come about?	Legacy of colonialism, institutional inertia, and dominant EU/global policy frameworks.	• Justice shaped by historic definitions of energy poverty (ch. 3). • Nuclear and tech conceived as 'inevitable' due to climate urgency (ch. 4). • Justice subsumed by decarbonisation agenda (ch. 4). • Implementation influenced by metrics, not social reality (chs. 2 and 3). • Justice literature fragmented and slow to operationalise (ch. 6).	From chapter 3 we take that defining and measuring energy poverty remains contested across academic and policy spheres. Academic debates span terms like *fuel poverty*, *energy poverty*, *vulnerability*, and *insecurity*, with the latter preferred for its broader socio-economic scope. EU policy defines energy poverty as multi-dimensional, tied to high energy costs, low income, and poor building performance. In Ireland, heavy import dependency and geopolitical shocks have exposed structural weaknesses in energy security and governance.

(continued)

Table 8.2 (continued)

WPR Questiony	*Summary of Representation*	*Core Themes Identified*	*Example of Supporting Representation arising from Chapter Analysis*
Q4: What is left unproblematic?	Technological fixes, power structures, global inequalities, and deeper justice dimensions.	• Technological dependency: No shift to sufficiency (ch. 2). • Displaced burdens: Marginalised communities must adapt (ch. 2). • Unequal access: Renters, Global South excluded (chs.3 and 8).	In chapter 2 several issues remain unaddressed. The paper relies on technological fixes rather than structural change. Responsibility is placed on communities, not industry or governance. Wider inequalities linked to energy poverty go unmeasured. The injustice of basic needs trade-offs is ignored. Health impacts are underreported. Global South harms and structural inequities are overlooked.
		• Justice silences: Gender, intergenerational, epistemic justices omitted (chs. 4, 5, 6, and 7). • Power not interrogated: Neoliberalism and elite control assumed (chs. 2, 5, and 6). • Global harm ignored: Extractivism in Global South unexamined (chs. 2, 4, and 8).	

Q5: What effects are produced?	Energy justice is reframed as a technocratic issue, sidelining systemic transformation in favour of expert-driven solutions.	• Technocratisation of justice: Operational toolkits replace politics (ch. 6) • Vulnerable subject creation: Passive framing (chs. 2, 3, and 5). • Academic authority: Grassroots voices minimised (ch. 6). • Sectoral fragmentation: Prevents systemic views (ch. 6). • Climate urgency > democracy: Deliberation sidelined (ch. 6). • Solution-orientation: 'Checklisting' makes radical change unthinkable (chs. 3 and 6). • Inclusion by exclusion: Early-career lens marginalises others (ch. 6).	In chapter 6 justice is framed as a technical issue, sidelining political transformation and centring principle alignment over power shifts. Policymakers are cast as problem-solvers; communities become objects of assessment, not co-creators. Methods may improve coherence but risk bureaucratising justice, privileging what's measurable and reproducing Global North logics. Policymakers and institutions gain legitimacy; marginalised communities with complex claims are left out.

(continued)

Table 8.2 (continued)

WPR Questiony	*Summary of Representation*	*Core Themes Identified*	*Example of Supporting Representation arising from Chapter Analysis*
Q6: How is the representation defended and/or disrupted?	Defended via legal, technical, and economic legitimacy; individualisation and procedural tokenism.[1] Disrupted via justice-centred narratives, structural critique, and decolonial resistance.	Representations defended by: Technocratic metrics and legal justifications; Individual responsibility framing; Tokenistic participation; Academic/policy elite discourse (chs. 3, 4, 5, and 6) Representations disrupted through: Democratic, equity-led narratives; Exposing silences and exclusion; Challenging growth and market norms; Decolonial knowledge systems; Grassroots mobilisation (chs. 2, 4, 5, 6, 7, and 8)	In chapter 8 the Green Deal's growth paradigm is fundamentally flawed and cannot be fixed through procedural or distributive tweaks. Recognition justice is easily sidelined when majorities benefit, leaving vulnerable or voiceless groups excluded. Relying on resource exports entrenches global inequalities, and weak corporate sustainability frameworks won't address this imbalance. A just transition requires structural transformation, not superficial adjustments.

Source (2025): Compiled by Sil Katrien Maslov-Lescrauwaet; Klaudia Szabelka; Rachel Walters; Tom Hambley and Päivi Tikkakoski.

[1] ***Procedural tokenism*** refers to symbolic or superficial forms of participation that give the appearance of inclusion without granting real influence, often used to legitimise predetermined decisions or maintain existing power structures.

energy justice is predominantly represented as a challenge of operationalising justice principles within existing systems (Q1). Unequal access to climate solutions, such as air conditioning, solar PV, or nuclear energy, is frequently highlighted, particularly in relation to those experiencing energy poverty. Implementation gaps are emphasised in these problem framings rather than the deeper structural drivers of injustice, such as entrenched economic inequality, centralised power in the energy system, or high-consumption lifestyles. Underlying these representations are assumptions that justice is achievable through technocratic, market-based, or incremental reforms (Q2), with policy metrics, stakeholder engagement, and regulatory adjustments being relied upon. A liberal, Global North epistemology is thereby reflected, in which procedural and distributive justice concerns are prioritised over recognitional or epistemic dimensions.

These representations are produced through a combination of institutional inertia, dominant policy frameworks, and historical legacies of fossil dependence and colonialism (Q3). As a result, key systemic issues, such as power imbalances, global extractivism, and the silencing of marginalised perspectives, are frequently left unproblematised (Q4). The effects of this framing are seen in the technocratisation and depoliticisation of justice, the construction of passive 'vulnerable subjects', the privileging of academic and expert knowledge, and the fragmentation of justice into sector-specific policy fixes (Q5). Nevertheless, while dominant narratives are legitimised through legal, economic, and technical discourse, disruptions to these framings are introduced in some chapters through justice-centred narratives, plural epistemologies, and critiques of market-based transitions (Q6). An emergent but constrained shift among early career scholars toward more structurally attuned and democratically engaged approaches to energy justice is thereby signalled.

8.2.6 Discussion

This critically informed, reflective analysis of ECR contributions through Bacchi's WPR framework reveals a pattern for how the problems that energy justice aims to address are represented and discussed. Despite an examination of chapters covering diverse contexts and technologies, spanning, for instance, nuclear energy governance[37] to solar PV policies,[38] from corporate regulation[39] to regional development,[40] there appears to be a consistent thematic pattern in these representations, whereby energy justice is primarily represented through the lens of technical, regulatory, or narrowly defined economic challenges. For example, energy poverty is framed primarily in terms of affordability rather than structural inequalities or regulatory challenges, or fundamentally politicised struggles over power and resources. Such insights have been brought to the forefront in recent scholarly publications (see e.g.[41,42])This pattern mirrors what sustainability transition scholars have long critiqued: the tendency to depoliticise systemic change by focusing on technical fixes rather than challenging power structures themselves.

The plurality of core energy justice concepts becomes particularly evident when presented by scholars from varied academic backgrounds. Whilst not a critique of the key strength of an interdisciplinary approach, evident both in the reviewed book and in energy justice scholarship more broadly,[43] it is nevertheless important to acknowledge the potential

[37] Basil, C. V. 2025. COP 28 Declaration on Nuclear Energy and the Energy Justice Framework: A Comparative Analysis, in: Phillips, A. and Kaschny, L. (Eds.), New Engagement Strategies for Energy Justice: Perspectives from the Next Generation. *Springer Nature Switzerland.*

[38] Spillane, J. 2025. Solar Rollout in Ireland: Revolution or Reinforcing Existing Social Inequity?, in: Phillips, A. and Kaschny, L. (Eds.), *New Engagement Strategies for Energy Justice: Perspectives from the Next Generation. Springer Nature Switzerland 2025.*

[39] Hazrati, M. and Annamayer, E. 2025. Evaluating EU Directives on Corporate Sustainability Reporting and Due Diligence Through the Lens of the 'JUST'Framework. In *New Engagement Strategies for Energy Justice: Perspectives from the Next Generation* (pp. 121–141). Cham: Springer Nature Switzerland.

[40] Obrenović, S. 2025. Never-Ending Transition: Energy Justice-Development Nexus in the Western Balkans. In *New Engagement Strategies for Energy Justice: Perspectives from the Next Generation* (pp. 107–120). Cham: Springer Nature Switzerland.

[41] Bouzarovski, S. *et al.* 2025. Reframing how we talk about 'energy poverty.' *Nature Energy*, *10*(7), 806–810. doi:10.1038/s41560-025-01794-w

[42] Dunlap, A. and Tornel, C. 2024. An insurrection in energy research: a dialogue between Carlos Tornel and Alexander Dunlap on energy justice, capitalist warfare & decolonization. *Globalizations*, *21*(1), 162–182. doi:10.1080/14747731.2023.2204543

[43] McCauley, D. *et al.* 2019. Energy justice in the transition to low carbon energy systems: Exploring key themes in interdisciplinary research. *Applied Energy*, 233–234, 916–921. doi:https://doi.org/10.1016/j.apenergy.2018.10.005

challenges posed by conceptual diversity. In chapter 2,[44] for example, energy justice is represented as a measurable shortcoming, which is addressable through improving the access to air conditioning and the provision of cooling shelters. The examination of Irish solar PV policy in chapter 3 constructs justice primarily as a problem of equitable access, wherein renters are framed as excluded, with policy intervention as a solution. Furthermore, how energy justice principles can be integrated into policies for a just transition in the Western Balkans[45] especially concerning the region's high reliance on coal, highlights initiatives like the Energy Community's Just Transition Forum, which brings together stakeholders for consultation and input on policies.[46] While it advocates for greater inclusion, the transition discourse appears to be dominated by Western European policy integration and hopes of political integration. Despite the intentions of transitioning away from past inequities, limited policy measures and inadequate institutional reforms fail to achieve the transformative systemic change that is necessary to address historical social injustices. It is seen how the findings across these diverse cases underscore a consistent pattern of reducing complex socio-political dynamics into more manageable, technical problems existing within prevailing institutional frameworks. Such framings risk serving existing governance arrangements by making injustice appear as implementation failures, rather than systemic design failures. This also echoes Ferguson's concept of the 'anti-politics machine'[47] inequality issues by converting them into technical problems, primarily strengthening bureaucratic power. In this section, we explore key themes from Table 2.2 using examples that emerged from our analysis of the chapters.

[44] Hoffmann, M. M., Kammen, D. M., Blechinger, P. 2025. Are Energy-Poor Communities in California More Vulnerable to Heat Wave-Related Health Burdens?, in: Phillips, A. and Kaschny, L. (Eds.), *New Engagement Strategies for Energy Justice: Perspectives from the Next Generation. Springer Nature Switzerland 2025.*

[45] Obrenović, S. 2025. Never-Ending Transition: Energy Justice-Development Nexus in the Western Balkans, in: Phillips, A. and Kaschny, L. (Eds.), *New Engagement Strategies for Energy Justice: Perspectives from the Next Generation. Springer Nature Switzerland 2025.*

[46] Obrenović, S. 2025. Never-Ending Transition: Energy Justice-Development Nexus in the Western Balkans, in: Phillips, A. and Kaschny, L. (Eds.), New Engagement Strategies for Energy Justice: Perspectives from the Next Generation. Springer Nature Switzerland 2025.

[47] Ferguson, J. 1994. *Anti-politics machine: Development, depoliticization, and bureaucratic power in Lesotho. University of Minnesota Press.*

8.3 Tendencies Towards Technocratic Framings of Justice or Policy-orientated Solutions

The analysis reveals how technical-solution-orientated framings suggest a tendency towards what we argue could be termed the 'technocratisation of justice.' In this process, the inherently political and contested nature of energy justice is represented as something that can be operationalised. The effects of this technocratisation are seen in how affected communities are represented across chapters. Californians vulnerable to heat become statistical categories defined by their inability to afford air conditioning, and Irish renters become policy subjects excluded from solar benefits due to tenure arrangements. In each case, what the analysis revealed is the construction of 'vulnerable subjects', or passive objects, that require expert protection, rather than recognition as active agents capable of defining their own needs and solutions.[48] Such a construction leads to the main emphasis being the legitimisation of expert intervention, while failing to question why these communities lack power in the first place. This framing of 'vulnerable subjects' also parallels Li's analysis[49] of how development discourse creates 'surplus populations'. Similarly, in this context, this refers to people who are problematised as being in need of expert management, rather than being recognised as agents of their own (energy) futures.

This pattern is reinforced by what the analysis suggests are types of knowledge being prioritised. Where policy evaluation, quantitative analysis and legal frameworks are viewed as more legitimate forms of evidence,[50] while grassroots knowledge, lived experiences,[51] and community-based solutions, which are at the heart of recognition justice, remain marginalised and underrepresented.[52] Moreover, the representation of the way in

[48] Jenkins, K., Sovacool, B. K., and McCauley, D. 2018. Humanizing sociotechnical transitions through energy justice: An ethical framework for global transformative change. *Energy policy*, *117*, 66–74.

[49] Li, T. M. 2017. After development: Surplus population and the politics of entitlement. *Development and Change 48*(6), 1247–1261.

[50] Sovacool, B. K. *et al.* 2017. New frontiers and conceptual frameworks for energy justice. *Energy policy*, *105*, 677–691.

[51] Willand, N., and Horne, R. 2018. 'They are grinding us into the ground'–The lived experience of (in) energy justice amongst low-income older households. *Applied Energy*, *226*, 61–70.

[52] Hanke, F., Guyet, R. and Feenstra, M. 2021. Do renewable energy communities deliver energy justice? Exploring insights from 71 European cases. *Energy Research & Social Science*, *80*, 102244.

which a just transition (using the energy justice framework) is structured as a 'toolkit',[53] with measurable principles exemplifies this rationalisation effect, whereby complex political struggles are converted into technical problems with measurable solutions.[54] While this could make justice appear more manageable and implementable to policymakers, it is argued that its political dimensions are simultaneously at risk of being stripped away. This potential depoliticisation may occur when justice is reduced or narrowed to technical compliance rather than addressing the power imbalances and structural inequalities that create injustices in the first place. As such, this risks transforming what should be contested political questions into seemingly neutral, administrative tasks, thereby potentially depoliticising the contested and transformative nature of justice claims. Essentially, the focus shifts from challenging existing power structures to optimising performance within those same structures. This pattern of emphasis on expert knowledge aligns with what is termed by decolonial scholars as the 'coloniality of knowledge',[55] whereby Western epistemologies have the systematic tendency to devalue Indigenous and local knowledge systems.

8.4 Fragmentation and the Politics of Scale

Malin and Ryder's concept of 'deep intersectionality'[56] reveals how environmental injustices emerge from interconnected systems of racial, gender, class, and other forms of domination. Still, the results reveal a pattern in terms of how the multi-dimensional issues surrounding energy (in)justice are fragmented into distinct sectoral problems. Each chapter focuses on different domains that represent energy justice as requiring specialised, sector-specific solutions. This fragmentation obscures what the analysis

[53] Kaschny, L. and Phillips, A. Lessons from New Engagement Strategies for Energy Justice, in: Phillips, A. and Kaschny, L. (Eds.), *New Engagement Strategies for Energy Justice: Perspectives from the Next Generation. Springer Nature Switzerland.*

[54] This concept draws on Weber's theory of rationalisation (Weber, M. 1978. *Economy and Society*. Berkeley: University of California Press) and aligns with Ferguson's 'anti-politics machine' discussed earlier, where development interventions convert political questions about power and inequality into technical problems that require expert management, thereby depoliticising structural issues.

[55] Quijano, A. 2000. Coloniality of power and Eurocentrism in Latin America. *International sociology*, *15*(2), 215–232.

[56] Malin, S. A., and Ryder, S. S. 2018. Developing deeply intersectional environmental justice scholarship. *Environmental Sociology*, *4*(1), 1–7.

suggests are broader political-economic patterns that underlie diverse manifestations of energy justice. While, for example, cases such as policy exclusions and heat vulnerability both reveal distributional injustices linked to housing tenure and income inequality, these connections remain largely unexplored. More specifically, due to such a fragmented approach, the attention given to systemic dimensions remains limited. The findings identify this fragmentation as a factor preventing the development of cross-cutting structural analyses that a justice approach demands, allowing the mere addressing of symptoms without confronting the systemic inequalities that produce energy injustice through multiple domains.

8.5 Energy Justice Becoming Normative

The results suggest that current trends in the use of the concept of energy justice risk becoming institutionalised in ways that diminish its potential. From the SDGs, pledges for net-zero emissions, and the commonly cited 'five pillars' of energy justice, a concerning shift could be conceptualised as energy justice being applied as a kind of 'justice checklist.' This trend tends to standardise or tokenise real-life struggles[57] into mere guidelines, checkboxes, or 'actionable' implementation frameworks for policymakers to act upon. Although while energy justice metrics may be useful for energy infrastructure economic policy modelling,[58] it is observed how a push towards defining *all* considerations of (in)justices using metrics and implementable, 'practical toolkits' reinforces this direction, despite good intentions (e.g. see [59]). Therefore, it is asserted that assumptions that energy justice needs to operate *within* existing structures, rather than challenging them, risk perpetuating inequalities.[60] In this process of institutionalisation, a process can be recognised which reflects what Escobar (2018) critiques as

[57] Dunlap, A., and Tornel, C. 2024. An insurrection in energy research: a dialogue between Carlos Tornel and Alexander Dunlap on energy justice, capitalist warfare & decolonization. *Globalizations, 21*(1), 162–182. doi:10.1080/14747731.2023.2204543

[58] Heffron, R. J., McCauley, D., and Sovacool, B. K. 2015. Resolving society's energy trilemma through the Energy Justice Metric. *Energy Policy*, 87, 168–176.

[59] Bouzarovski, S. *et al.* 2025. Reframing how we talk about 'energy poverty.' *Nature Energy, 10*(7), 806–810. doi:10.1038/s41560-025-01794-w

[60] Dunlap, A. and Tornel, C. (2024). An insurrection in energy research: a dialogue between Carlos Tornel and Alexander Dunlap on energy justice, capitalist warfare & decolonization. *Globalizations, 21*(1), 162–182. doi:10.1080/14747731.2023.2204543

'ontological design'.[61] Here, practices reproduce modernist assumptions about progress while marginalising alternative ways of being. Escobar's framework of 'autonomous design',[62] instead, calls for approaches that favour more collaborative and place-based approaches. The support for this argument becomes particularly strong when delving into how inconsistencies and disagreements are addressed and managed. Rather than embracing the complexity and plurality through which issues of justice are understood, considering all diverse perspectives – including those that are often overlooked, marginalised, made 'invisible', or silenced – it is seen that the reviewed chapters focus on finding practical or technical resolutions that tend to steer clear of deeply engaging in political debates.[63] This kind of 'rush', which Le Merle[64] referred to as the tension between 'climate emergency and time-consuming democratic processes', to integrate energy justice into policy-relevant discussions has inadvertently led to what can be viewed as a 'flattening' of justice discussions, suggesting technical fixes rather than addressing broader issues related to sovereignty, agency, recognition, redistribution, and representation may become less visible when primarily framed as problems of efficiency, access, and consultation. Such a kind of reduction risks diluting the essence of justice claims and obscuring the political complexities and power dynamic struggles that green energy transitions – and many other sustainability projects[65] – inevitably entail. In doing so, scholarship aiming to advance energy justice may inadvertently contribute to the reproduction of top-down, expert-driven approaches and misrecognition.

To synthesise the insights generated through the application of Bacchi's WPR structure, an Iceberg Model (Fig. 8.1) was developed to visually represent the layered nature of how energy justice is constructed within ECR contributions. To structure and communicate these findings more

[61] Escobar, A., and Maffei, S. 2021. What are pluriversal politics and ontological designing? Interview with Arturo Escobar. *diid—disegno industriale industrial design*, (75), 12–12.

[62] Escobar, A. 2018. *Designs for the pluriverse: Radical interdependence, autonomy, and the making of worlds*. Duke University Press.

[63] Sovacool, B. K. *et al.* 2020. Sociotechnical agendas: Reviewing future directions for energy and climate research. *Energy Research & Social Science*, *70*, 101617.

[64] Kaschny, L. and Phillips, A. 2025. Lessons from New Engagement Strategies for Energy Justice, in: Phillips, A. and Kaschny, L. (Eds.), *New Engagement Strategies for Energy Justice: Perspectives from the Next Generation. Springer Nature Switzerland*

[65] McGowan, K., and Antadze, N. 2023. Recognizing the dark side of sustainability transitions. *Journal of Environmental Studies and Sciences*, *13*(2), 344–349.

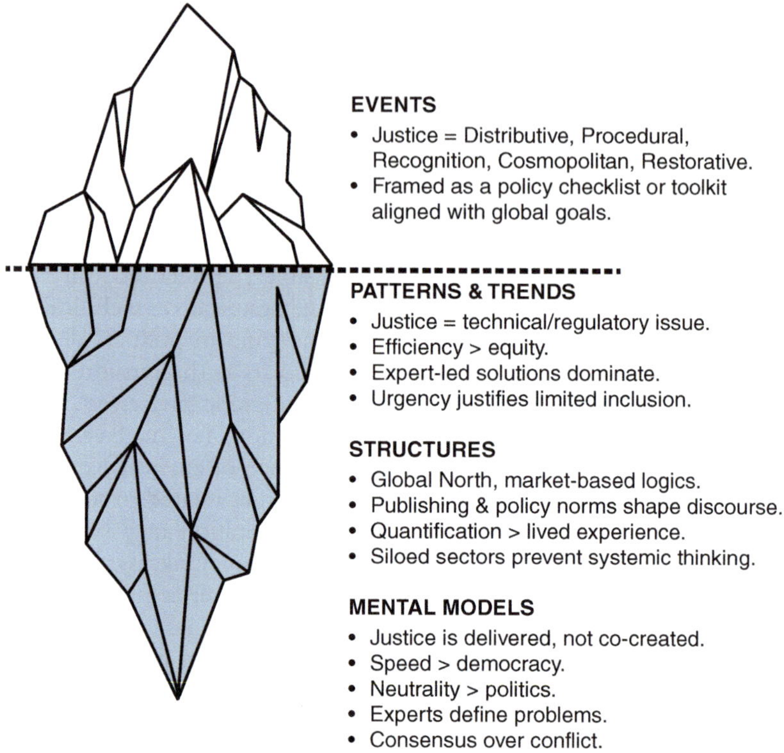

Fig. 8.1 The Iceberg Model of Energy Justice Representations. Source: Adapted from Hemingway's Iceberg Theory. Hemingway, E. 1932. Death in the Afternoon (Scribner's 1932) 192–193 (explaining the 'Iceberg Theory').

effectively, the dominant themes were clustered, drawing on the metaphor of an iceberg to illustrate the relationship between surface-level discourses and deeper, less visible systemic and ideological factors.

At the surface level ('Events'), the dominant framing of energy justice is identified as a normative checklist – typically structured around the five tenets (distributive, procedural, recognition, cosmopolitan, and restorative justice). These justice pillars are frequently mobilised in alignment with global agendas (e.g., SDGs, net-zero transitions), giving the appearance of comprehensiveness while masking deeper contestations. The next level, 'Patterns and Trends', reflects the recurring tendencies across chapters to frame justice in technical or regulatory terms, privileging efficiency over equity and expert-led problem-solving over participatory approaches. These patterns reveal how representations tend to depoliticise justice by positioning it as a deliverable solution rather than a contested process. 'Structures' capture the institutional and systemic influences – such as Global North epistemologies, publishing norms, and technocratic paradigms – that shape how justice is framed and constrain what can be said. These structures normalise quantification and sectoral fragmentation, which limit systemic or intersectional critique. Finally, at the deepest level, 'Mental Models',[66] the often-unspoken assumptions and ideological commitments that govern the discursive space, are identified. These include beliefs that justice is something to be delivered rather than co-created, that speed trumps deliberation, and that consensus is preferable to political conflict. Naturally, the iceberg is depicted in its context (i.e., the ocean). The ocean illustrates a call to approach energy (in)justice not only by considering one of the five pillars but instead by approaching related challenges with a holistic view that also acknowledges diverse backgrounds and intersectionality. These underlying logics reveal how epistemic authority and normative neutrality reinforce dominant representations and reinforce misrecognition justice. The Iceberg Model thus acts as a critical-visual complement to our WPR analysis, making visible the multi-layered ways in which justice is discursively constructed, stabilised, and constrained in early career scholarship.

8.6 Moving Forward

The analysis points towards several concrete directions for change. First, the consistent pattern of academic expertise overshadowing community-driven knowledge suggests the need for collaborative approaches. Rather than studying vulnerable populations, energy justice research could engage with what Conde[67] terms 'Activism Mobilizing Science' (AMS). AMS refers to collaborative processes where activists, communities, and researchers

[66] Here we conceptualized the importance of the role of mental models although a formal analysis was not conducted, evidence of such insights is available from key research undertaken in this area by van den Broek. *et al.* 2025. Expert and citizen perceptions of the drivers of the energy transition: A mental model approach. *Journal of Cleaner Production*, 494, 144949.

[67] Conde, M. 2014. Activism mobilising science. *Ecological economics*, *105*, 67–77.

co-produce knowledge to address energy injustices while challenging unequal power dynamics. Second, the issue of fragmentation suggests the need for more systemic analytical frameworks. Instead of approaching energy justice as sectoral problems requiring specialised solutions, research work towards developing analyses that trace connections across scales without reducing its complexity into manageable variables. Lastly, the temporal urgency effects revealed by the analysis suggest taking into consideration balancing this temporal effect with 'Slow Science'[68] – a research approach that is open to uncertainty and conflict, rather than aiming for quick policy closure. As such, a call is made for energy justice research in which epistemological reflexivity and critical reflection can be benefited from, rather than a descriptive approach in which solutions are formulated to an assumed problem (cf. What is being left unproblematic?). Scholars are strongly recommended to apply the WPR scheme to their own work.

8.7 Reflection on the WPR Framework

Academic publishing stipulations, academic publishing norms, and institutional pressures are acknowledged as factors that may contribute to shaping how ECRs write about justice.[69] Additionally, work undertaken by ECRs is, by necessity, narrow in scope to gain nuanced perspectives. Thus, the WPR framework offers a welcome approach to identify and challenge the wider structural issues that result in societal inequalities. That is to assist with the identification of what is *not* said, what is taken for granted, or that which may be missing. Additionally, to interrogate not only the argument presented (e.g., in the chapters) but also how it governs the space of possible thought and action. Its strength lies in revealing how even well-intentioned, methodologically rigorous scholarship can contribute to the stabilisation of dominant paradigms, such as liberal universalism or rationalist policymaking.

It is important to be reflexive and acknowledge our own positionality as part of this analysis process. As ECRs, we acknowledge the limitations of this work and provide recommendations for future research directions. As such, the various perspectives and diverse range of socio-spatial case

[68] Stengers, I. 2011. Another science is possible!. *A plea for slow science*, 2011–2012.

[69] Dunlap, A., and Tornel, C. 2024. An insurrection in energy research: a dialogue between Carlos Tornel and Alexander Dunlap on energy justice, capitalist warfare & decolonization. *Globalizations, 21*(1), 162–182. doi:10.1080/14747731.2023.2204543

studies this book offers are welcomed. Also, the research topic presentations provide interdisciplinary and cross-disciplinary scholarly perspectives, reflecting what energy justice aims to encompass and achieve in practice. The very nature of energy transitions acknowledges the complexities these socio-technical transitions encompass, and 'the problems' are often politically constructed with material consequences. Energy justice is often implicitly or explicitly stated in some capacity, but challenges remain for implementation in practice. Thus, the different epistemological orientations (e.g., critical theory, feminist theory, postcolonial critiques, etc.) the various chapters provide are an important contribution to this dialogue. The plurality of energy justice enables dominant paradigms to be identified and dismantled, as necessary. It also enhances understanding of where academia can go further, where efforts may be fruitlessly focused, and where challenges may be more effectively pursued in the future.

8.8 Conclusion

This chapter has applied Carol Bacchi's 'What's the Problem Represented to be?' analysis scheme to critically examine how ECRs represent the problems which energy justice aims to address in their scholarship. We found that applying the WPR framework to energy justice scholarship proves to be an appropriate and useful tool to add critical reflexivity in the field of energy justice. The framework allows researchers to reveal underlying assumptions, identify silences, and examine the effects of certain representations that energy justice aims to address, since the field inherently deals with contested concepts such as fairness, power and transformation. We therefore recommend energy justice scholars to apply the WPR framework to their own work and to the work of others to reflect on the dominant paradigms and representations within the field. By applying the WPR, we found normative assumptions embedded within the scholarship. The analysis revealed that, despite the diversity of examined contexts, the problems that energy justice aims to address tend to be represented as technical or regulatory deficiencies instead of fundamental systemic challenges. This normative representation grounds energy justice in assumptions that risk reducing complex injustices. The implications for this pattern extend beyond the small-scale level or individual impacts.

As ECRs ourselves, a call is made for energy justice research that primarily acknowledges complexity and conflict, rather than prioritising its management. This requires moving beyond treating justice issues through toolkit-like approaches that invoke a tenet when appropriate towards approaches that explore the deeper issues beneath those tenets to centre community knowledge, trace connections across scales, and remain open to uncertainty, rather than rushing towards policy closure. In this warming world, icebergs will inevitably melt, and their hidden depths will become visible as they dissolve into the one and the same surrounding ocean. Long submerged beneath technocratic surfaces, the structural foundations of injustice will demand our attention as the urgency of climate action exposes their fundamental role in shaping whose voices are heard, whose knowledge counts, and whose futures matter.

CHAPTER 9

Building the Just Transition and Circular Economy Future

Alicia Phillips and Laura Kaschny

Abstract The journey toward a sustainable future requires much more than technological innovations or economic remodelling. A profound and distinct reimagining of how we exist, live, work and relate to one another and the planet is needed. Global perspectives on the circular economy in the Just Transition herald the change we need to adopt. While a just transition calls on the need to ensure that no one is left behind as we shift from carbon-intensive systems, the circular economy offers the structural and material pathways needed to bring this vision to life. By converging on a shared vision and embedding justice at the centre, the core circular economy strategies of reform, reuse and reduce become the one ethical, practical and people-centred approach needed. Now with the climate urgency emergency, the question is no longer about whether we can afford to transition, but whether we can afford not to.

A. Phillips
UMR TREE, CNRS, Université de Pau et des Pays de l'Adour, Pau, France
e-mail: alicia.phillips@univ-pau.fr

L. Kaschny
Faculty of Law, Maastricht University, Maastricht, The Netherlands
e-mail: laura.kaschny@maastrichtuniversity.nl

A. Phillips, L. Kaschny (eds.), *Global Perspectives on the Circular Economy in the Just Transition*, Just Transitions,
https://doi.org/10.1007/978-3-032-18342-2_9

Keywords Sustainability • Sustainable Future • Circular economy • Just transition • Justice

9.1 Systemic Transformation Principles Towards a Just Transition

Circular economy strategies are already presenting opportunities for local and national stakeholders looking to boost social and economic development in ways that also tackle environmental issues.[1] Indeed, while the just transition has emerged over time as a guiding framework for achieving sustainability without leaving vulnerable communities behind, a central aspect of this vision should be the circular economy. When integrating the principles of energy justice into the equation to breathe life into ensuring the transition is indeed just, a more intent-orientated plan is born. The book has taken us through the ways the just transition is being managed, the latent issues that arise and how infusing circular economy strategies can aid in mitigating these issues, primarily by placing justice at the centre of policies going forward. While the just transition is a complex topic with many layers and perspectives that differ between countries and demographic groups,[2] the book has presented some of the main issues that are being tackled today and proffered ways to analyse and process these issues.

This book has introduced the circular economy as a conceptual framework that advances a just transition through three interrelated strategies: (1) reform, (2) reuse, and (3) reduce. Reform aims to address the systemic structures, institutions and policies that shape society, just transitions and root causes of unsustainability. Reuse emphasises product design and processes to maximise resource efficiency, extend material lifecycles, and promote recycling, refurbishment and responsible consumption. Reduce focuses on minimising environmental impacts, lowering emissions, and

[1] Goodwin Brown, E. and Glutting, J. 2024. It's time to get specific on how circular economy strategies can service just transitions. Materia Rinnovabile. Available at: https://www.renewablematter.eu/en/Its-time-to-get-specific-on-how-circular-economy-strategies-can-service-just-transitions (Accessed: 6 November 2025).

[2] SLYCAN Trust. 2021. Challenges and Opportunities for a Just Transition. Available at: https://weadapt.org/wp-content/uploads/2023/05/issues_paper_on_challenges_and_opportunities_for_a_just_transition_0_0.pdf (Accessed: 6 November 2025).

reducing waste through strategies such as energy efficiency and renewable energy integration. These three pillars provide the conceptual foundation, which is further explored and applied through the lens of the book's chapters, offering concrete illustrations of how circular economy strategies can facilitate socially and environmentally just transitions across diverse contexts, cultures and realities.

9.2 The Case for Reform

The global community is facing an unprecedented convergence of crises involving climate change, ecological degradation, deepening inequality and social injustice. Challenges like these expose the unsustainability of the current economic and social systems that are rooted in unequal power relations, uncertain energy futures and inefficient spatial planning. Confronting these issues is no easy task, and for societies to embark on a truly just transition, these issues must be systemically approached with deliberate processes by reforming, restructuring and reorganising to transform society.

Firstly, global efforts need to address the drivers of unjust transitions in dominant economic models and uneven power relations.[3] Gender is fundamental to the success and failure of sustainability transitions and their justness because gender is fundamental to power relations and inequalities.[4] Further, while the economic and environmental potentials of the transition are explicitly reflected in the circular economy, its relationship with the social dimension has been inadequately explored both theoretically and practically.[5] Naturally, since the transition is rooted in societal changes that involve paradigm shifts in values, cultures, laws, societal structures and other social norms, the social dimension is also pivotal to the circular economy's success.[6]

[3] Newell, P. *et al.* 2023. Mind the gap: The global governance of just transitions. *Global Policy*, *14*(3): 425–437. https://doi.org/10.1111/1758-5899.13236

[4] Avelino, F. *et al.* 2024. Just Sustainability Transitions: Politics, power and prefiguration in transformative change toward justice and sustainability. *Annual Review of Environment and Resources*, *49*. https://doi.org/10.1146/annurev-environ-112321-081722

[5] Liu, K. 2024. Circular economy and the separated yet inseparable social dimension: Views from European circular city experts. *Sustainable Production and Consumption*, *51*: 474–483. https://doi.org/10.1016/j.spc.2024.09.016

[6] Ibid.

Women have been at the forefront of climate action for decades, leading movements, resistance, research and initiatives, including testing innovative tools and hosting participatory spaces, yet structural issues are holding back women's roles in sectoral transitions.[7] *Chapter 2* advocates for the inclusion of these silenced voices, notably those of indigenous women of colour in the Global South who have been at the frontline of advancing energy just futures. By examining the contributions of the *MAB* in Brazil and *WoMin* on the African continent, the chapter examines critically the tangible and innovative ways these women-led organisations have championed stronger spaces for women in societies where their voices are silenced. If the energy transition is to be truly just, it must centre on the voices, needs and leadership of women and under-represented communities.[8] *Chapter 2* is evidence of the impact these two groups have had in shaping the energy futures of their communities, with both being central to highlighting the impacts of energy activities as they relate to the constructions of dams and extractive industries, respectively, calling on the need for justice in the process. Notably, this chapter is evidence that respecting the priorities and needs of communities is essential to ensuring well-informed, inclusive and accountable decision-making on the energy transition.[9]

Secondly, *Chapter 3* has highlighted the need to correct uncertain energy futures by restructuring energy systems. By exploring Russia's invasion of Ukraine in 2022 and the ensuing energy crisis in the EU, a reshaping of the EU energy strategy has taken place since it has forced Europe to speed up the energy transition.[10] Trade patterns for oil and natural gas have also shifted dramatically since Russia's invasion, as governments look

[7] Martinelli, C. 2025. A just future needs gender and climate justice. Climate Action Network. Available at: https://caneurope.org/a-just-future-needs-gender-and-climate-justice/ (Accessed: 6 November 2025).

[8] Vazi, B. and Laan, T. 2025. Why Gender inclusion must drive the Just energy transition. International Institute for Sustainable Development. Available at: https://www.iisd.org/articles/insight/why-gender-inclusion-must-drive-just-energy-transition (Accessed: 6 November 2025).

[9] Extractive Industries Transparency Initiative. 2023. Engaging communities in a just transition. Available at: https://eiti.org/sites/default/files/2023-10/EITI_Report_Engaging%20communities%20in%20a%20just%20transition.pdf (Accessed: 6 November 2025).

[10] Holman, R. and Lefort, J. 2024. Invested in Renewables: The only way forward. European Investment Bank. Available at: https://www.eib.org/en/essays/europe-energy-transition-renewable (Accessed: 6 November 2025).

to strengthen their energy security.[11] The chapter posits an examination of the EU's nuclear governance framework through the lens of energy justice and proposes a future-proof regulatory framework that integrates legal clarity, technological innovation and democratic legitimacy to ensure a just energy transition in the EU.

Historically, the three leading concerns of nuclear energy were nuclear accidents, radioactive nuclear waste disposal and potential health effects[12] but in 2023 nuclear energy returned to the centre of EU energy debates with the Declaration to Triple Nuclear Energy Capacity by 2050 signed at the United Nations Climate Change Conference in Dubai.[13] By examining the case of nuclear power in the EU alongside energy justice principles, *Chapter 3* offers a comprehensive analysis of assessing the concerns that underlie the EU's nuclear policy, which entails different forms of inequality that occur throughout the entire nuclear energy lifecycle. The chapter proffers that while nuclear power can play a role in responding to the rising demand for secure energy in Europe, the governance system fails to address several injustices which occur. Generally, therefore, a deeper reconfiguration of this governance system is needed to ensure it is equitable and in line with the just transition. While energy equity and justice have become priority considerations for policymakers, to ensure that energy equity is incorporated into actual decisions and analysis, it is necessary to design, use, and continually improve energy equity metrics.[14] The chapter encourages a reflection for EU policymakers on restructuring the complex balance of energy security, meeting climate objectives and addressing justice considerations.

[11] International Energy Agency. Russia's war on Ukraine: Analysing the impacts on energy markets and energy security. Available at: https://www.iea.org/topics/russias-war-on-ukraine (Accessed: 6 November 2025).

[12] Ho, J. *et al.* 2014. Perceived environmental and health risks of nuclear energy in Taiwan after Fukushima nuclear disaster. *Environmental International*, *73*: 295–303. https://doi.org/10.1016/j.envint.2014.08.007

[13] World Nuclear Association. 2025. Landmark Ministerial Declaration puts nuclear energy at the heart of action on climate change. Available at: https://world-nuclear.org/net-zero-nuclear/news/landmark-ministerial-declaration-puts-nuclear-energy-at-the-heart-of-action-on-climate-change (Accessed: 6 November 2025).

[14] Baker, E. *et al.* 2023. Metrics for decision-making in energy justice. *Annual Reviews of Environment and Resources*, *48*: 737–460. https://doi.org/10.1146/annurev-environ-112621-063400

Thirdly, *Chapter 4* seeks to examine how inefficient spatial planning is being corrected by urban regeneration and renewable energy projects in Italy and Brazil. Although urbanisation is a cornerstone of sustainable development and has led to a massive demand for energy to power economic activity, expand infrastructure and deliver municipal services, it can lead to unsustainable and costly energy consumption patterns.[15] In efforts to make cities more sustainable, energy services are therefore a priority, especially in the context of the Sustainable Development Goals, particularly Goal 11.[16] By examining how urban regeneration can promote energy justice and ensure fair and inclusive access to energy in the cities of Roma in Italy and Brasília in Brazil, the chapter adopts an interdisciplinary approach that integrates legal and political analysis with empirical research to investigate the relationship between urban regeneration and energy justice.

The implementation of the energy justice lens here is useful since justice considerations are becoming increasingly important in spatial planning research and practice.[17] Reorganisation here therefore extends beyond policy and economics into the social and cultural fabric of society and requires community empowerment via renewable energy cooperatives and participatory planning, making energy communities key tools for tackling energy poverty, strengthening civic participation and enhancing social cohesion. Spatial planning is therefore crucial for energy planning due to the spatial dimension of both energy demand and energy availability, and when it comes to cities, the need to ensure a secure supply of energy as it relates to the clean energy transition is a priority given their dense settlements.[18] Further, by presenting how energy communities in both cases function, the chapter presents valuable lessons on the importance of renewable energy communities as key enablers of socio-urban regeneration.

[15] Energy Sector Management Assistance Program. Energy Efficient cities. Available at: https://www.esmap.org/node/57881 (Accessed: 7 November 2025).

[16] United Nations. Department of Economic and Social Affairs. Sustainable Development Goal 11: Make cities and human settlements inclusive, safe, resilient and sustainable. Available at: https://sdgs.un.org/goals/goal11 (Accessed: 7 November 2025).

[17] Rot, C. *et al.* 2025. Is it just planning? Unpacking markers of justice in spatial planning literature. *Journal of Planning Literature*. https://doi.org/10.1177/08854122251328259

[18] Integrated Energy, Climate and Spatial planning to enable local and regional authorities to effectively implement their plan. 2.1. Renewable energy. Available at: https://ieecp.org/inplanpractice/focus-topics-for-climate-action/renewable-energy/ (Accessed: 7 November 2025).

9.3 The Reuse Claim

In an era filled with rapid consumerism and escalating environmental pressures, the concept of reuse has emerged as a cornerstone of sustainable development. Sustainability rests on interlinked pillars of environmental protection, social well-being and economic viability, and reuse plays a distinct role by encouraging circular economies, minimising waste reduction and reducing extraction. The chapters in this section seek to examine the concepts of consumerism via the fast fashion crisis in Bangladesh and the growing demand for minerals and metals as the global decarbonisation process moves forward.

In *Chapter 5* the examination of the fast fashion crisis in Bangladesh by the examination of the Rana Plaza incident is used to highlight the limitations of narrowly focused reforms and the need for holistic sustainability approaches that address not just labour rights but also environmental degradation, overconsumption and global economic inequalities. Particularly, it draws attention to the demand-side patterns in the Global North that sustain the fast fashion model, connecting unchecked consumption and the constant pursuit of low-cost garments to both environmental harm and labour exploitation. As such, the chapter argues that a just transition must be anticipatory, participatory and globally coordinated to integrate social and ecological justice concerns across the entire supply chain. By utilising the energy justice lens and introducing the degrowth lens into the investigation, the chapter illustrates that a framework built on the principles of degrowth can help reconcile several issues. Here, the concept of reuse is relied upon to extend the lifecycle of garments and reduce material waste and ultimately to offer a more practical approach to reducing environmental pressures while aligning consumption and production with more just and sustainable practices.

Chapter 6 sought to elucidate the dilemma faced in the mining sector in Latin America between maximising the economic benefits of mining while simultaneously safeguarding the environment and upholding human rights at the same time. It situates this dilemma within the context of the global energy transition by emphasising how rising international demand for critical minerals such as copper and lithium intensifies pressures on extractive practices. The chapter particularly argues that this increase in demand reinforces historical patterns of economic dependency and socio-environmental inequities. By looking at the judgement of the Inter-American Court of

Human Rights in the case of *Inhabitants of La Oroya v Peru*[19] the chapter utilised the restorative justice principle to show how it could contribute to the remediation and prevention of environmental harm and public health issues arising from extractive activities in the region. The chapter was principal in providing insights on how a restorative justice framework could be used to offer a transformative perspective to the dilemma being faced in the region. The concepts of reuse and resource efficiency are emphasised as strategies to reduce environmental impacts, and mitigate extractive pressures with the objective of ensuring that the burdens and benefits of mining are more equitably distributed across the value chain.

9.4 The Urgency to Reduce GHG Emissions

Carbon emission reduction is the most critical means of addressing climate change, and with the increasingly prominent impact of carbon emissions on the environment, society and the economy, carbon emission reduction has become a global priority.[20] The concept of 'reduce' calls for minimising resource consumption while avoiding waste across the value chain. As such, it forms a complementary strategy to the energy transition through renewable energy deployment and low-carbon technologies.

In *Chapter 7* the authors provided an in-depth procedural justice analysis of the implementation of cross-border green hydrogen initiatives between the EU and Tunisia and the risks that such projects may pose to ensuring an inclusive and equitable energy transition in Tunisia. In doing so the chapter examines the historical relationship of the EU with Tunisia and the absence of adequate procedural justice safeguards in some of the public and private legal arrangements underpinning the SoutH2 Corridor project. The chapter situates these hydrogen projects within the broader context of resource use, highlighting the efficient use of materials and infrastructure as essential to prevent the repetition of extractive and inequitable practices within energy systems. Fundamentally, while hydrogen has gained momentum as a potential solution for decarbonizing

[19] Inter-American Court of Human Rights. Case of the Inhabitants of La Oroya v Peru. Available at: https://www.corteidh.or.cr/docs/casos/articulos/seriec_511_ing.pdf (Accessed: 7 November 2025).

[20] Wang, N. *et al.* 2024. Synergies between the circular economy and carbon emission reduction. *Science of the Total Environment*, 951. https://doi.org/10.1016/j.scitotenv.2024.175603

several equity implications exist,[21] but by acknowledging these dynamics, Europe can play an instrumental role in shaping equitable frameworks that promote justice for both local communities in partner countries and the broader global equity.[22] By integrating resource efficiency into project design and operational strategies, this chapter emphasises decarbonisation objectives that form an opportunity to connect the EU and Tunisia in such a way that it fosters a more collaborative and equitable energy transition between partner and host countries.

9.5 Analysing the Use of the Critical Reflective Framework Tool

Chapter 8, though outside the book's focus on the circular economy and just transition, provided a noteworthy analysis of early career research. By utilising the 'What is the problem represented to be' (WPR) approach, the authors sought to highlight the underlying assumptions, identify silences and examine the effects of certain representations that energy justice aims to address. The chapter further highlighted a call for more energy justice research to better acknowledge complexity and conflict rather than its management by utilisation of the WPR approach. However, one can also argue that the utilisation of such a tool allows for research to be not only vast and unabounding but unsystemic in dealing with a specifically identifiable problem or issue. Instinctively, while the authors have advocated for the use of the WPR approach in response to a toolkit, one can argue that the toolkit approach is a more targeted approach to address problems in society. There is no doubt that the WPR is also a tool and could provide improved communication to different stakeholders.

9.6 Final Thoughts

As we conclude this book, the chapters of this book have demonstrated that the concepts of reform, reuse and reduce cannot be considered merely as isolated strategies but are mutually reinforcing pathways fostering systemic

[21] Dejonghe, M. and de Graaf, T. 2025. Green Colonialism or green transformation? The equity implications of clean hydrogen trade. *Political Geography*, 120. https://doi.org/10.1016/j.polgeo.2025.103338

[22] Patonia, A. 2025. Green hydrogen and its unspoken challenges for energy justice. *Applied Energy*, 377, Part C. https://doi.org/10.1016/j.apenergy.2024.124674

and holistic change. 'Reform' emphasises the need for restructuring societal, economic, and institutional systems with the aim of addressing the root causes of environmental degradation and social inequality, both from a local to a more global stage. The concept of 'reuse' reflects the imperative of extending the lifecycle of resources across a wide variety of materials, including garments, minerals or energy infrastructure. As such, it promotes social equity through the reduction of waste and the alleviation of pressures on ecosystems. 'Reduce' emphasises the need of minimising consumption across sectors. The case of emissions also demonstrates that climate measures are inseparable from justice concerns, not least through a fair distribution of resources across geographies. Collectively, these lessons reveal that circular economy strategies go beyond technical fixes but provide a conceptual lens through which human impact, equity across the value chain and a just transition should be revised.

We look forward to the task ahead and hope that the research offered in this text encourages the need for a holistic transformation that is ethical, practical and people-centred. It demands a new mindset, one that places intrinsic value on stewardship over consumption, resilience overgrowth and collective well-being over individual gain. Overall, by aligning justice with the need to reform, reuse and reduce, we have the opportunity to create a world that thrives with a respect for planetary boundaries. Finally, linking the just transition with the circular economy provides a framework that protects the livelihoods of future generations.

Bibliography

Ackerly, B.A. and True, J. (2019). *Doing Feminist Research in Political and Social Science* (2nd edn, Macmillan International 2020)

Abe, O. and Azubike, V. (2024). (Re)Examining the Intersection between Energy Justice and Energy Transition in Africa. *Journal of Energy & Natural Resources Law*, 42 (3): 279–99. https://doi.org/10.1080/02646811.2024.2356988

Abousahl, S., *et al.* 2021. *Technical assessment of nuclear energy with respect to the 'do no significant harm' criteria of Regulation (EU) 2020/852 ('Taxonomy Regulation')*. Publications Office of the European Union

Afrose, M. 2024. EU Due-Diligence Directive for Corporate Sustainability: Consequences for Bangladesh. Bangladesh Foreign Trade Institute. Available at: https://www.bfti.org.bd/blog-details/11 (Accessed: 20 August 2025)

Ahmed, W. et al. 2024. Circular economy: Extending end-of-life strategies. *Sustainable Production and Consumption*, 51: 67–78. https://doi.org/10.1016/j.spc.2024.09.003.

Akhter, S. 2014. Endless Misery of Nimble Fingers: The Rana Plaza Disaster. *AJWS* 20(1): 137–47. https://doi.org/10.1080/12259276.2014.11666176

Albright, E. A. 2011. Policy Change and Learning in Response to Extreme Flood Events in Hungary: An Advocacy Coalition Approach. *Policy Studies Journal*, 39(3): 485–511.Available at: https://doi.org/10.1111/j.1541-0072.2011.00418.x

Alam, H. et al. 2024. International Circular Economy Policy Trends - Implications for Bangladesh's Circular Garment and Textile Policy Development. Available at:

A. Phillips, L. Kaschny (eds.), *Global Perspectives on the Circular Economy in the Just Transition*, Just Transitions,
https://doi.org/10.1007/978-3-032-18342-2

https://www.switchtocircular.eu/sites/default/files/publications/International%20circular%20economy%20policy%20trends_implications%20for%20Bangladesh_pub_web.pdf

Albert, M. J. 2024. Navigating the Polycrisis: Mapping the Futures of Capitalism and the Earth. The MIT Press.

Allouche, B. and Abdeljalil, A. 2020. L'enseignement supérieur en Tunisie: A-t-on sacrifié la qualité face aux pressions quantitatives? In: Revue marocaine de la pensée contemporaine, 2020, n° 5. Available at: https://archive-ouverte.unige.ch/unige:171798. (Accessed: 20 October 2025)

Amara, T. 2024. Tunisia assembly votes to strip court of electoral authority, days before presidential vote. *Reuters*. Available at: https://www.reuters.com/world/africa/tunisia-assembly-votes-electoral-bill-nine-days-before-poll-opposition-calls-2024-09-27/. (Accessed: 27 July 2025)

Ammar, E., & Ammar, S. 2024. Green hydrogen in Tunisia: A new mechanism of plunder and exploitation. Transnational Institute. Available at: https://www.tni.org/en/article/green-hydrogen-in-tunisia;

Ammar, I. 2022. Towards a just energy transition in Tunisia Transnational Institute. Available at:https://www.tni.org/files/2023-01/energy_democracy_tunesia_web.pdf. (Accessed: 20 October 2025)

Ammar, S. 2024. Resisting the new green colonialism. Transnational Institute. Available at: https://www.tni.org/en/article/resisting-the-new-green-colonialism (Accessed: 15 May 2025)

Amnesty International. 2024. Garment workers must receive rights-based compensation and justice immediately. Available at: https://www.amnesty.org/en/latest/news/2024/05/bangladesh-garment-workers-must-receive-rights-based-compensation-and-justice-immediately/ (Accessed: 20 August 2025)

Antmann, M. 2025. Capitalism and resource extraction: A Marxist value-theory approach to Latin American socio-environmental conflicts. *The Extractive Industries and Society*, 22. https://doi.org/10.1016/j.exis.2025.101617

Apergi, M. *et al.* 2024. An Energy Justice Index for the Energy Transition in the Global South. *Renewable and Sustainable Energy Reviews*, 192. https://doi.org/10.1016/j.rser.2023.114238;

Arboleda, M. 2020. Planetary Mine: Territories of Extraction under Late Capitalism. Verso

Arias-Loyola, M. *et al.* 2022. Beyond the Resource Curse: The Redistributive Challenge of Sustainable Resource-Led Development in Australia, Chile and Zambia. *The Extractive Industries and Society*, 11. https://doi.org/10.1016/j.exis.2022.101084.

Arora, S. *et al.* (2025). Decolonising Innovation for Pluriversal Justice in Sustainability Transitions. Social Science Research Network.

Article 194(2) TFEU

Article 2 (16) in Directive (EU) 2018/2001 read together with Article 2 (11), Directive (EU) 2019/944

Article 16 in Directive (EU) 2019/944;
Article 267 TFEU read along with the Recommendations to national courts and tribunals in relation to the initiation of preliminary ruling proceedings, OJC C/2024/6008, 9.10.2024l
Article 2 (11) (a) in Directive (EU) 2019/944 of the European Parliament and of the Council of 5 June 2019 on common rules for the internal market for electricity and amending Directive 2012/27/EU, OJL 158, 14.6.2019
Article 2(36), RED III
Articles 27.1 (a) (ii), (d), 27.6, and 29a of RED III
Article 1.2 of the European Parliament and the Council Regulation (EU) 2022/869 of 30 May 2022 on guidelines for trans-European energy infrastructure, amending Regulations (EC) No 715/2009, (EU) 2019/942 and (EU) 2019/943 and Directive 2009/73/EC and (EU) 2019/944, and repealing Regulation (EU) No 347/2013 OJ L 152/45 ("TEN-E Regulation").
Articles 18 and 19 of the TEN-E Regulation and Article 5(2) of Regulation (EU) 2021/1153 of the European Parliament and of the Council of 7 July 2021 establishing the Connecting Europe Facility and repealing Regulations (EU) No 1316/2013 and (EU) No 283/2014 (OJ L 249, 14.7.2021.
Article 4.2 of TEN-E Regulation
Article 1.2 TEN-E Regulation
Article 2, 3.5 & 21 of the Consolidated Treaty on European Union of 26 October 2012 OJ C 326
Article 191 of the Consolidated Treaty on the functioning of the European Union of 26 October 2012 OJ C 326
Art 5 .4 (a) of the TEN-E Regulation
Article 22 in Directive (EU) 2018/2001
Avelino, F. et al. 2024. Just Sustainability Transitions: Politics, power and prefiguration in transformative change toward justice and sustainability. *Annual Review of Environment and Resources*, 49. https://doi.org/10.1146/annurev-environ-112321-081722
Bacchi, C., 2012. Introducing the 'What's the Problem Represented to be' approach. *Engaging with Carol Bacchi: Strategic interventions and exchanges*, pp.21–4.
Bacchi, C. 2009. Analysing Policy: What's the problem represented to be? Introduction. *Pearson Australia*. pp. Ix-xxii
Bacchi, C., & Goodwin, S. 2016. Introduction. In Poststructural policy analysis: A guide to practice (pp. 3–12). Springer.
Baker, E. et al. 2023. Metrics for decision-making in energy justice. Annual Reviews of Environment and Resources, 48: 737–460. https://doi.org/10.1146/annurev-environ-112621-063400.
Bamber, P. et al. 2023. Innovation and Competitiveness in the Copper-Mining GVC: Developing Local Suppliers in Peru. *Industrial and Corporate Change*, 33 (4): 940, 941. https://doi.org/10.1093/icc/dtad033
Bangladesh Labour Act, 2006 (Act No. XLII of 2006), enacted on 11 October 2006

Bahishti, Adam. 2022. The Vital role of research methodology in addressing the research questions. International Journal of Methodology, 1. https://doi.org/10.21467/ijm.1.1.2016.29

Balletto, G. *et al.* 2022. More Circular City in the Energy and Ecological Transition: A Methodological Approach to Sustainable Urban Regeneration. *Sustainability*:14 (22). https://doi.org/10.3390/su142214995.

Barja-Martínez, S., *et al.* 2025. Local energy communities optimization considering cost and greenhouse gases minimization. https://arxiv.org/abs/2504.20973

Barua, D. 2025. The Transfer from Fast Fashion to Slow Fashion: Challenges and Opportunities for Bangladesh's RMG Sector. *International Journal on Science and Technology*: 16(3)

Barnard, M. 2021. Nuclear Generation Has Huge and Hidden Security Costs. The Future Is Electric. *Medium*. Available at: https://medium.com/the-future-is-electric/nuclear-generation-has-huge-and-hidden-security-costs-e0a11264bbac. (Accessed: 9 June 2025).

Basil, C.V. 2025. COP 28 Declaration on Nuclear Energy and the Energy Justice Framework: A Comparative Analysis, in: Phillips, A., Kaschny, L. (Eds.), New Engagement Strategies for Energy Justice: Perspectives from the Next Generation. *Springer Nature Switzerland*

Bastida, A.E. 2020. The Law and Governance of Mining and Minerals: A Global Perspective. Hart Publishing, 5.

Baynernet is completely owned and controlled by German government institutions and municipalities. Available at: https://www.bayernets.de/en/about-us/who-we-are/shareholders. (Accessed: 20 October 2025)

Bell, S.E. *et al.* (2020). Toward Feminist Energy Systems: Why Adding Women and Solar Panels Is Not Enough☆. *Energy Research & Social Science*, 68. https://doi.org/10.1016/j.erss.2020.101557;

Bellinkx, V. *et al.* 2022. Addressing Climate Change through International Human Rights Law: From (Extra)Territoriality to Common Concern of Humankind. *Transnational E*nvironmental *Law*,11 (1): 69–93. doi:10.1017/S204710252100011X

Bezerra, P. et al. 2022. The multidimensionality of energy poverty in Brazil: A historical analysis. *Energy Policy, 171*. https://doi.org/10.1016/j.enpol.2022.113268

Bhambra, G. K. 2022. A Decolonial Project for Europe. *JCMS: Journal of Common Market Studies* 60(2): 229–44. https://doi.org/10.1111/jcms.13310

Birkland, T. A. 1998. Focusing Events, Mobilization, and Agenda Setting. *Journal of Public Policy*, 18(1): 53–74. Available at: https://www.jstor.org/stable/4007601;

Birkland, T. A. 2009. Disasters, Lessons Learned, and Fantasy Documents. *Journal of Contingencies and Crisis Management* 17(3): 146–56. https://doi.org/10.1111/j.1468-5973.2009.00575.x

Bohdanowicz, Z. *et al.* 2022. Europe Becomes Pro-nuclear? Drivers of public support for nuclear energy in six EU countries after the energy crisis of 2022. 2025. *Energy Policy* 200: https://doi.org/10.1016/j.enpol.2025.114528.

Bohdanowicz, Z., *et al.* 2023. Support for Nuclear Power and Pro-environmental Attitudes: The cases of Germany and Poland. *Energy Policy,* 177. https://doi.org/10.1016/j.enpol.2023.113578

Bonavolontà, F., *et al.* 2025. Real-Time Monitoring of Energy Contributions in Renewable Energy Communities Through an IoT Measurement System. *Sensors,* 25(5). https://doi.org/10.3390/s25051402.

Bouzarovski, S. and Simcock, N. 2017. Spatializing energy justice. Energy Policy, *107,* pp.640–8.

Bouzarovski, S., Cedano-Villavicencio, K. G., Delina, L. L., Martiskainen, M., Reames, T. G., & Simcock, N. (2025). Reframing how we talk about 'energy poverty.' Nature Energy, 10(7), 806–10. https://doi.org/10.1038/s41560-025-01794-w.

Brand, U. and Wissen, M. 2021. The Imperial Mode of Living: Everyday Life and the Ecological Crisis of Capitalism. London: Verso

Brasil de Fato. 2025. 'MAB Realiza Atos Em Defesa de Comunidades Atingidas Por Barragens Pelo Brasil.' Available at: https://www.brasildefato.com.br/2025/03/14/mab-realiza-atos-em-defesa-de-comunidades-atingidas-por-barragens-pelo-brasil/. (Accessed: 29 June 2025);

Brasil de Fato. 2017. 'De Atingidos Por Barragens a Defensores Dos Rios, Da Água e Da Vida.' Available at: https://www.brasildefato.com.br/2017/03/16/de-atingidos-por-barragens-a-defensores-dos-rios-da-agua-e-da-vida/ (Accessed: 10 June 2025)

Brook, B. *et al.* 2014. Why Nuclear Energy is Sustainable and Has to be part of the Energy Mix. *Sustainable Materials and Technologies,* 1–2: 8–16. https://doi.org/10.1016/j.susmat.2014.11.001.

Brown, A., and P. Börkey. 2024. Extended producer responsibility in the garments sector. *OECD Environment Working Papers,* 253. https://doi.org/10.1787/8ee5adb2-en

Buckingham, S. 2015. Ecofeminism. *International Encyclopedia of the Social & Behavioral Sciences.* Elsevier.

Building a World-Ship: A Decolonial Approach to Just Energy Transitions. *Human Geography.* https://doi.org/10.1177/19427786251326605;

Bury, J. 2005. Mining Mountains: Neoliberalism, Land Tenure, Livelihoods, and the New Peruvian Mining Industry in Cajamarca. *Environment and Planning A: Economy and Space,* 37(2): 221, 224. https://doi.org/10.1068/a371

Butler, C., Parkhill, K.A. and Luzecka, P. 2018. Rethinking energy demand governance: Exploring impact beyond 'energy'policy. *Energy research & social science, 36,* pp.70–8.

Cacciuttolo, C. et al. 2023. Socio-Environmental Risks Linked with Mine Tailings Chemical Composition: Promoting Responsible and Safe Mine Tailings

Management Considering Copper and Gold Mining Experiences from Chile and Peru. *Toxics*, 11 (5). https://doi.org/10.3390/toxics11050462

Campagna, L., *et al.* 2024. Renewable energy communities and mitigation of energy poverty: Instruments for policymakers and community managers. *Sustainable Energy, Grids and Networks*, 39. https://doi.org/10.1016/j.segan.2024.101471.

Caramizaru, A., & Uihlein, A. 2020. Energy communities: An overview of energy and social innovation (EUR 30083 EN). Luxembourg: Publications Office of the European Union.

Caroline Omari Lichuma, "(Laws) Made in the 'First World': A TWAIL Critique of the Use of Domestic Legislation to Extraterritorially Regulate Global Value Chains" (2021) 81 Zeitschrift für ausländisches öffentliches Recht und Völkerrecht / Heidelberg Journal of International Law 497.

Case of La Oroya Population v. Peru. (Judgment of November 27, 2023, on Preliminary Objections, Merits, Reparations and Costs) Inter-American Court of Human Rights Ser C No 511.

Case C-664/15 Protect Natur-, Arten- und Landschaftsschutz Umweltorganisation v Bezirkshauptmannschaft Gmünd,ECLI:EU:C:2017:98

Cassa Depositi e Prestiti S.p.A (2023)

Case T-215/23, ClientEarth and Others v Commission, OJ C 235, 3.7.2023

Centre Tricontinental (CETRI). 2024. Green Hydrogen: A New Mechanism of Green Colonialism? Available at: https://www.cetri.be/Green-hydrogen-A-new-mechanism-of

Centobelli, P. et al. 2022. Slowing the fast fashion industry: An all-round perspective. *Current Opinion in Green and Sustainable Chemistry*, 38. https://doi.org/10.1016/j.cogsc.2022.100684

CER Le Vele. 2025. Available at: https://www.cerlevele.it. (Accessed: 31 July 2025)

CER Le Vele. Analisi dei Dati Energetici 2024. Available at: https://www.cerlevele.it. (Accessed: 31 July 2025).

Chagnon, C. et al. 2022. From Extractivism to Global Extractivism: The Evolution of an Organizing Concept. *The Journal of Peasant Studies*, 49 (4): 760–92. https://doi.org/10.1080/03066150.2022.2069015

Choudhury, M and Haque, C.M. 2024. Disaster management policy changes in Bangladesh: Drivers and factors of a shift from reactive to proactive approach. *Environmental Policy and Governance*, 34(5):445–62. https://doi.org/10.1002/eet.2094

Cigna, L. et al. 2023. Varieties of Just Transition? Eco-Social Policy Approaches at the International Level. *Social Policy and Society*, 22(4): 730–46. https://doi.org/10.1017/S1474746423000192

Claar, S. 2022. Green Colonialism in the European Green Deal: Continuities of Dependency and the Relationship of Forces between Europe and Africa. *Culture, Practice & Europeanization*, 7(2): 262–74. DOI:10.5771/2566-7742-2022-2-262

Clean Clothes Campaign. Rana Plaza. Clean Clothes Campaign. Available at: https://cleanclothes.org/campaigns/past/rana-plaza. (Accessed: 20 August 2025)

Clean Hydrogen Partnership, "The European hydrogen market landscape," November 2024. Available at https://observatory.clean-hydrogen.europa.eu/sites/default/files/2024-11/The%20European%20hydrogen%20market%20landscape_November%202024.pdf (Accessed:11 August 2025)

Climate Action Network Europe. 2024. Myth Buster: Nuclear energy is a dangerous distraction. Available at: https://caneurope.org/myth-buster-nuclear-energy/. (Accessed: 23 April 2025)

Cocco, A. 2022. The Role of Energy Communities in the Energy Transition. *The Italian Law Journal*, *2*, 509–25. DOI 10.23815/2421-2156.ITALJ

Coglianese, C. and Nash, J. 2001. *Regulating from the Inside: Can Environmental Management Systems Achieve Policy Goals?* RFF Press

Crowther, A., Petrova, S. and Evans, J. 2023. Between vision and implementation: the exclusionary disjuncture of domestic heat decarbonisation in Greater Manchester. *Local Environment*, *28*(8), pp.1045–61.

Couldry, N. and Mejias, U.A. (2021). The Decolonial Turn in Data and Technology Research: What Is at Stake and Where Is It Heading? *Information, Communication & Society*, *26(4)*:786–802. https://doi.org/10.1080/1369118X.2021.1986102

Coglianese, C. and Nash, J. 2001. Regulating from the Inside: Can Environmental Management Systems Achieve Policy Goals? RFF Press

Communication COM(2019) 640 final, The European Green Deal, (EU Green Deal); European Commission. 2021.

Communication COM(2021) 550 final, Fit for 55': delivering the EU's 2030 Climate Targets on the way to climate neutrality.

Comune di Roma. 2025. Regolamento CER Assemblea Capitolina. Available at: https://www.comune.roma.it/webresources/cms/documents/Mun11_Regolamento_CER_AssembleaCapitolina_1.pdf. (Accessed: 20 August 2025)

Commission Delegated Regulation (EU) 2021/2139 of 4 June 2021, *OJ L 442, 9.12.2021;* Commission Delegated Regulation (EU) 2021/2178 of 6 July 2021, *OJ L 443, 10.12.2021*

Commission Delegated Regulation (EU) 2022/1214 of 9 March 2022 (covering specific gas and nuclear activities), *OJ L 188, 15.7.2022*

Commission Delegated Regulation (EU) 2023/2486 of 27 June 2023, plus related amendment act (EU) 2023/2485, *OJ L, 2023/2486, 21.11.2023,*A Commission Delegated Act amending the Disclosures, Climate and Environmental Delegated Acts adopted 4 July 2025. The changes are set to apply from January 1, 2026, with the option to apply them to the 2025 financial year.

Convention on Access to Information, Public Participation in Decision-Making and Access to Justice in Environmental Matters (Aarhus Convention), Aarhus, 25 June 1998.

Conde, M. 2014. Activism mobilising science. *Ecological economics*, *105*, 67–77.

Corporate Europe Observatory. 2025. We say no to the South H2 Corridor: A statement from 87 CSOs. Available at: https://corporateeurope.org/en/corporateeurope.org/en/2025/03/CSOs-reject-South-H2-corridor. (Accessed: 20 October 2025)

Corvellec, H. Stowell, Alison, F. and Johansson, N. 2022. Critiques of the circular economy. *Journal of Industrial Ecology,* 26(2): 421–32. https://doi.org/10.1111/jiec.13187

Cotton, M. 2021. Nuclear Power and Environmental Justice: The case for political equality. In: Kaiser, A., Lehtonen, M., Meyer, J.H. & Rubio, M. (eds), Engaging the Atom: The history of nuclear energy and society in Europe from the 1950s to the present, Morgantown: West Virginia University Press.

Council Directive 2009/71/Euratom of 25 June 2009 establishing a Community framework for the nuclear safety of nuclear installations, OJL 172, 2.7.2009.

Council Directive 2011/70/Euratom of 19 July 2011 establishing a Community framework for the responsible and safe management of spent nuclear fuel and radioactive waste, OJ L 199, 2.8.2011.

Council Directive 2013/59/Euratom of 5 December 2013, laying down basic safety standards for protection against the dangers arising from exposure to ionising radiation, OJL 13, 17.1.2014

Crabtree, J. 2024. Peru and the Search for a Development Model That Works. In: Amann, E. and Figueiredo, P.A. (Eds). Innovation, Competitiveness, and Development in Latin America.Oxford University Press New York, 394.

Crabtree, J. and Durand, F. 2017. Peru: Elite Power and Political Capture. Zed Books.

Crenshaw, K. (1991). Mapping the Margins: Intersectionality, Identity Politics, and Violence against Women of Color. *Stanford Law Review,* 43 (6): 1241–99. https://doi.org/10.2307/1229039.

Cummings, S. *et al.* (2023). Doing Epistemic Justice in Sustainable Development: Applying the Philosophical Concept of Epistemic Injustice to the Real World. *Sustainable Development,* 31(3):1965–77. https://doi.org/10.1002/sd.2497

D'Alpaos, C., & Andreolli, F. 2021. Renewable energy communities: The challenge for new policy and regulatory frameworks design. *New metropolitan perspectives: knowledge dynamics and innovation-driven policies towards urban and regional transition, 2*, 500–509. Di Lorenzo, G., *et al.* 2021. Innovative power-sharing model for buildings and energy communities. *Renewable Energy, 172*: 1087–102. https://doi.org/10.1016/j.renene.2021.03.063

Delgado, M., & Rufin, C. 2025. Energy transition policies in the Global North and the Global South: Comparing the US and Brazil. Social Science Research Network. https://dx.doi.org/10.2139/ssrn.5117055.

Delegated Regulation. Available at: https://energy.ec.europa.eu/document/download/944b96b9-4efd-44a3-bbfe-45b752b0b55f_en (Accessed: 10 October 2025).

D'Alisa, G. et al. 2015. Degrowth: A Vocabulary for a New Era. Routledge; Fletcher, Kate, and Mathilda Tham. 201; Earth Logic: Fashion Action Research Plan. Available at: https://katefletcher.com/wp-content/uploads/2019/10/Earth-Logic-plan-FINAL.pdf (Accessed: 20 August 2025)

Dargent, E. et al. 2017. Greater State Capacity, Lesser Stateness: Lessons from the Peruvian Commodity Boom.

Demaria,F.*et al.* 2013. What is Degrowth? From an Activist Slogan to a Social Movement. *Environmental Values* 22(2): 191–215. https://doi.org/10.2307/23460978

D'Onofrio, R., & Trusiani, E. 2022. The Future of the City in the Name of Proximity: A New Perspective for the Urban Regeneration of Council Housing Suburbs in Italy after the Pandemic. Sustainability, 14(3). https://doi.org/10.3390/su14031252.

De Cicco, M. C., & Ferra Júnior, A. R. 2022. Fundamental duty to sustainable development in the Brazilian Federal Constitution of 1988 and solar energy (SDG 7) as an energy policy in the current Brazilian conjuncture. In: L. Ruggeri & F. Pascucci (Eds.), Prosumerism and energy communities: Expanding concepts in a global perspective (pp. 75–83). Scholarly Society Publisher of Wien.

De Freitas, A. (2019). Decolonial Epistemologies for Energy Planning in Brazil. PhD Thesis, University of Westminster School of Law. https://doi.org/10.34737/v3497

Dejonghe, M. and Van de Graaf, T. 2025. Green colonialism or green transformation? The equity implications of clean hydrogen trade. *Political Geography*, 120. https://doi.org/10.1016/j.polgeo.2025.103338.

DeLeo, R. A. 2017. Anticipatory policymaking in global venues: Policy change, adaptation, and the UNFCCC. *Futures*, 92: 39–47. https://doi.org/10.1016/j.futures.2016.09.001

de Oliveira, F.A. 2024. Dependency and Crisis in Brazil and Argentina: A Critique of Market and State Utopias. University of Pittsburgh Press.

da Silva, E. C. M. et al. 2024. The Role of Renewable Energies in Combating Poverty in Brazil: A Systematic Review. *Sustainability*, *16*(13). https://doi.org/10.3390/su16135584

de Souza, A. *et al.* 2023. A comparison study of grid impact of photovoltaic installations in Brazil according to Normative Resolution 482 and Federal law 14.300. Social Science Research Network. https://dx.doi.org/10.2139/ssrn.4309470.

Deutsche Gesellschaft für Internationale Zusammenarbeit (GIZ). 2022. Establishing a Bavarian-Tunisian Technology and Innovation Hub for Green Hydrogen. Available at: https://www.giz.de/en/downloads/giz2022-en-bavaria-tunisia.pdf. (Accessed: 20 October 2025)

Dindar, S. (2025). A Systematic Review of Urban Regeneration's Impact on Sustainable Transport: Traffic Dynamics, Policy Responses, and Environmental Implications. *Sustainable Development*. https://doi.org/10.1002/sd.70007

Directive 2003/4/EC of the European Parliament and of the Council of 28 January 2003 on public access to environmental information and repealing Council Directive 90/313/EEC, OJL 41, 14.2.2003;

Directive 2011/92/EU of the European Parliament and of the Council of 13 December 2011 on the assessment of the effects of certain public and private projects on the environment, as amended by Directive 2014/52/EU, OJL 26, 28.1.2012

Directive 2014/52/EU of the European Parliament and of the Council of 16 April 2014 amending Directive 2011/92/EU on the assessment of the effects of certain public and private projects on the environment, OJL 124, 25.4.2014

Directive (EU) 2018/2001 of the European Parliament and of the Council of 11 December 2018 on the promotion of the use of energy from renewable sources (recast) (RED II), as amended until 2024, OJL 328, 21.12.2018

Directive (EU) 2024/1711 of the European Parliament and of the Council of 13 June 2024 amending Directives (EU) 2018/2001 and (EU) 2019/944 as regards improving the EU's electricity market design, OJ L, 2024/1711, 26.6.2024.

Directive (EU) 2019/944 of the European Parliament and of the Council of 5 June 2019 on common rules for the internal market for electricity and amending Directive 2012/27/EU (as amended until 2024), OJL 158, 14.6.2019

Dougherty, M.L. 2016. From Global Peripheries to the Earth's Core The New Extraction in Latin America In: Deonandan, K. and Dougherty, M.L. (Eds), Mining in Latin America: Critical approaches to the new extraction. Routledge. 5.

Droubi, S., Heffron, R.J. and McCauley, D. 2022. A critical review of energy democracy: A failure to deliver justice? *Energy Research & Social Science*, *86*, p.102444

Dudka, S. and Adriano, D.C. 1997. Environmental Impacts of Metal Ore Mining and Processing: A Review. *Journal of Environmental Quality*, 26 (3): 590– 602. https://doi.org/10.2134/jeq1997.00472425002600030003x.

Dunlap, A., & Tornel, C. (2024). An insurrection in energy research: a dialogue between Carlos Tornel and Alexander Dunlap on energy justice, capitalist warfare & decolonization. Globalizations, 21(1), 162–82. doi:10.1080/14747731.2023.2204543

Durante, F. et al. 2021. Extraction and Extractivisms: Definitions and Concepts. In: Shapiro, J. and McNeish, J.A. (Eds) Our Extractive Age: Expressions of Violence and Resistance. Routledge

Dynge, M. and Cali, U. (2025). Distributive Energy Justice in Local Electricity Markets: Assessing the Performance of Fairness Indicators. *Applied Energy*, 384. https://doi.org/10.1016/j.apenergy.2025.125463

Ecorys. 2013. Final Report - Trade Sustainability Impact Assessment in support of negotiations of a DCFTA between the EU and Tunisia. (Rotterdam). Pg 177 – 186.Available at: https://doi.org/10.13140/RG.2.1.1886.0400. (Accessed: 20 October 2025)

Edwards, R., 2004. Nuclear disarray as Europe pushes east. New Scientist. Available at: https://www.newscientist.com/article/mg18224451-000-nuclear-disarray-as-europe-pushes-east/. (Accessed: 11 August 2025).

Eeckhout, Lieven. 2023. Towards sustainable computer architecture: A holistic approach. In: Duranton, M. et al. (Eds) *HiPEAC Vision 2023*;

El Amine, Y. 2025. Just Transition Green Bridge: Tunisian Case Study. Arab Reform Initiative. Available at: https://www.arab-reform.net/publication/just-transition-green-bridge-tunisian-case-study/ (Accessed: 20 October 2025)

Ellen MacArthur Foundation. 2017. A new textiles economy: Redesigning fashion's future. Available at: https://content.ellenmacarthurfoundation.org/m/6d5071bb8a5f05a2/original/A-New-Textiles-Economy-Redesigning-fashions-future.pdf (Accessed: 20 August 2025)

Elmqvist, T., et al. (eds.). 2015. Urbanization, biodiversity and ecosystem services: Challenges and opportunities. Springer. https://doi.org/10.1007/978-94-007-7088-1

Energy Sector Management Assistance Program. Energy Efficient cities. Available at: https://www.esmap.org/node/57881. (Accessed: 7 November 2025)

Engstrom, P. and Pérez, E. 2025. Confronting Extractivism. In: Behrend, J. and Whitehead, L. (Eds). Mounting Pressures on the Rule of Law Routledge

Ernst and Young (EY). 2024. The True Power of Small Modular Reactors on the Road to A Sustainable Energy Future: Unveiling key opportunities and challenges Paris: EY. Available at: https://www.ey.com/content/dam/ey-unified-site/ey-com/fr-fr/insights/energy-resources/documents/ey-etude-nuclear-energy-smr-pov-20240322.pdf. (Accessed: 22 May 2025)

Escobar, A., & Maffei, S. 2021. What are pluriversal politics and ontological design? Interview with Arturo Escobar. *diid—disegno industriale industrial design*, (75), 12–12.

Escobar, A. 2018. *Designs for the pluriverse: Radical interdependence, autonomy, and the making of worlds.* Duke University Press.

Espinoza-Guillen, J.A. *et al.* 2023. Immission Levels and Identification of Sulfur Dioxide Sources in La Oroya City, Peruvian Andes. *Environment, Development and Sustainability*, 25: 12843–72. https://doi.org/10.1007/s10668-022-02592-0.

EU Hydrogen Strategy; Recital 13 of the Preamble to the European Commission. 2024. Directive (EU) 2024/1788 on common rules for the internal markets for renewable gas, natural gas and hydrogen, amending Directive (EU) 2023/1791 and repealing Directive 2009/73/EC, OJ L 15.7.2024 (Renewable Gases and Hydrogen Directive). Available at: https://eur-lex.europa.eu/eli/dir/2024/1788/oj/eng. (Accessed on: 20 October 2025)

EU Hydrogen Strategy. Available at: https://eur-lex.europa.eu/eli/dir/2024/1788/oj/eng. (Accessed on: 20 October 2025)

Euro-Mediterranean Agreement establishing an Association between the European Communities and their Member States, of the one part, and the Republic of

Tunisia, of the other part, *Official Journal of the European Union*, L 97, 30 March 1998.

EU Neighbours South. 2024. Tunisia: PPP, ELMED and the ambitious MoU Toward Green Hydrogen and Renewable Energy. Available at: https://south.euneighbours.eu/news/tunisia-ppp-elmed-and-the-ambitious-mou-toward-green-hydrogen-and-renewable-energy/ (Accessed: 15 May 2025).

European Parliament. Euratom Treaty. Available at: https://www.europarl.europa.eu/about-parliament/en/in-the-past/the-parliament-and-the-treaties/euratom-treaty. (Accessed: 20 October 2025)

Euractiv. 2025. MEPs endorse rollback of corporate sustainability rules on EPP terms. Available at: https://www.euractiv.com/news/meps-endorse-rollback-of-corporate-sustainability-rules-on-epp-terms/. (Accessed: 20 October 2025).

European commission. 2025. Hydrogen PCI/PMI East Meeting. Available at: https://webcast.ec.europa.eu/hydrogen-pcipmi-east-meeting-2025-04-14 (Accessed: 10 October 2025).

European Commission. 2023. Memorandum of Understanding on a strategic and global partnership between the European Union and Tunisia. Available at: https://ec.europa.eu/commission/presscorner/detail/en/ip_23_3887. (Accessed: 23 May 2025).

European Commission. 2025. New European Bauhaus: beautiful, sustainable, together. Available at: https://new-european-bauhaus.europa.eu/index_en. (Accessed: 21 August 2025).

European Commission. 2022. First Circular Economy Package including Regulation proposal on Sustainable Products and Sustainable Textiles Strategy. Available at: https://ec.europa.eu/newsroom/env/items/742027/en. (Accessed: 20 August 2025;

European Commission. 2021. Joint Communication, JOIN(2021) 30 final, The Global Gateway.

European Commission. 2023. Euratom research in action and opportunities for Europe: EU strategic autonomy and the future energy systems: EU Small Modular Reactors (SMRs) Declaration. Publications Office of the European Union. https://data.europa.eu/doi/10.2777/911262.

European Commission. 2020. Regulation (EU) 2020/852. Available at: https://eur-lex.europa.eu/eli/reg/2020/852/oj/eng. (Accessed: 20 October 2025)

European Commission, Directorate-General for Research and Innovation. 2023. Declaration on EU SMR 2030: The role of Research, Innovation, Education and Training in the safety of Small Modular Reactors (SMRs) in the European Union. Luxembourg: Publications Office of the European Union.

European Commission. Directorate-General for Energy. 2024. Energy poverty. European Commission. Available at: https://energy.ec.europa.eu/topics/markets-and-consumers/energy-consumers-and-prosumers/energy-poverty_en?utm_source=chatgpt.com. (Accessed: 19 Aug 2025).

European Parliament. 2023. Small Modular Reactors: Opportunities and Challenges for the EU, Policy Department for Industry, Research and Energy. Available at: https://energy.ec.europa.eu/topics/nuclear-energy/small-modular-reactors_en. (Accessed:12 May 2025).

European Commission. 2022. Communication COM(2022) 230 final, REPowerEU Plan (REPowerEU Plan)

European Commission. 2025. Communication COM(2025) 85 final, The Clean Industrial Deal: A joint roadmap for competitiveness and decarbonisation (Clean Industrial Deal)

European Commission. 2025. Communication COM(2025) 79 final, Action Plan for Affordable Energy

European Commission.2020. European Commission. 2022. Communication COMCommunication COM(2020) 301 final, a hydrogen strategy for climate-neutral Europe (EU Hydrogen Strategy)

European Commission. 2020. Communication COM(2020) 299 final, Powering a climate-neutral economy: An EU Strategy for Energy System Integration (EU Energy System Integration Strategy).

European Commission. 2021. Joint Communication, JOIN(2021) 30 final, The Global Gateway (The Global Gateway).

European Commission. 2023. Delegated Regulation (EU) 2023/1184 of 10 February 2023 supplementing Directive (EU) 2018/2001 of the European Parliament and of the Council by establishing a Union methodology setting out detailed rules for the production of renewable liquid and gaseous transport fuels of non-biological origin. [2023] OJ L 157/11.

European Commission. 2023. Energy Communities and Energy Poverty. Available at: https://publications.jrc.ec.europa.eu/repository/handle/JRC134832. (Accessed: 20 October 2025)

European Commission. 2023. Delegated Regulation (EU) 2023/1185 of 10 February 2023 supplementing Directive (EU) 2018/2001 of the European Parliament and of the Council by establishing a minimum threshold for greenhouse gas emissions savings of recycled carbon fuels and by specifying a methodology for assessing greenhouse gas emissions savings from renewable liquid and gaseous transport fuels of non-biological origin and from recycled carbon fuels. [2023] OJ L 157/20.

European Commission. 2023. Memorandum of Understanding on a strategic and global partnership between the European Union and Tunisia. Available at: https://ec.europa.eu/commission/presscorner/detail/en/ip_23_3887 (Accessed: 23 May 2025)

European Commission. 2017. Residential prosumers in the European Energy Union. Publications Office of the European Union. Available at: https://commission.europa.eu/system/files/2017-11/study-residential-prosumers-energy-union_en.pdf. (Accessed: 2 Sep 2025)

European Commission. 2016. Clean Energy for All Europeans (COM (2016) 860 final). Available at: https://eur-lex.europa.eu/legal-content/EN/TXT/HTML/?uri=CELEX:52016DC0860. (Accessed: 20 October 2025).

European Think Tanks Group (ETTG). A new chapter in the EU–North Africa Partnership on Climate and Energy. 2024. Available at: https://ettg.eu/eu-north-africa-partnership-climate-energy/. (Accessed: 20 October 2025)

European Parliament, Directive (EU) 2024/1788 on common rules for the internal markets for renewable gas, natural gas and hydrogen, amending Directive (EU) 2023/1791 and repealing Directive 2009/73/EC, OJ L, 2024/1788, 15.7.2024 (RGH Directive).

European Parliament. 2023. UN Climate Change Conference 2023 in Dubai, United Arab Emirates (COP28) (2023/263). Available at: https://www.europarl.europa.eu/doceo/document/TA-9-2023-0407_EN.html. (Accessed: 20 October 2025)

European Parliament. Directive (EU) 2019/944. Available at: https://eur-lex.europa.eu/eli/dir/2019/944/oj/eng. (Accessed: 20 October 2025)

European Parliament and Council. 2024. Directive (EU) 2024/1760 on corporate sustainability due diligence and amending Directive (EU) 2019/1937 and Regulation (EU) 2023/2859, OJ L 2024/1760, .05.07.2024 [CSDDD]

European Parliament and Council. 2022. Directive (EU) 2022/2464 amending Regulation (EU) No 537/2014, Directive 2004/109/EC, Directive 2006/43/EC and Directive 2013/34/EU, as regards corporate sustainability reporting, OJ L 322, 16.12.2022 [CSRD].

European Parliament. 2021. Regulation (EU) 2021/1767. Available at: https://eur-lex.europa.eu/eli/reg/2021/1767/oj/eng. (Accessed: 20 October 2025)

European Parliament resolution of 14 March 2024 on the adoption of the special measure in favour of Tunisia for 2023 (2024/2573(RSP)), P9_TA(2024)0175, 14 March 2024 https://www.europarl.europa.eu/doceo/document/TA-9-2024-0175_EN.html

European Hydrogen Backbone. 2022. Five Hydrogen Supply Corridors for Europe in 2030. Available at: https://ehb.eu/files/downloads/EHB-Supply-corridor-presentation-Full-version.pdf/ (Accessed:13 August 2025).

European Parliament and Council. 2023. Directive (EU) 2023/2413 amending Directive (EU) 2018/2001, Regulation (EU) 2018/1999 and Directive 98/70/EC as regards the promotion of energy from renewable sources, and repealing Council Directive (EU) 2015/652, OJ L 31.10.2023 (RED III).

European Parliament. 2023. Regulation (EU) 2023/on the use of renewable and low-carbon fuels in maritime transport, and amending Directive 2009/16/ECEU, OJ L 234, 22.9.2023 (FuelEU Maritime).

European Parliament and the Council Regulation (EU) 2022/869 of 30 May 2022 on guidelines for trans-European energy infrastructure, amending Regulations (EC) No 715/2009, (EU) 2019/942 and (EU) 2019/943 and

Directive 2009/73/EC and (EU) 2019/944, and repealing Regulation (EU) No 347/2013 OJ L 152/45 ("TEN-E Regulation").

European Parliament. European Parliamentary Research Service. *Nuclear energy in the European Union* (2023), 1–9.

European Union. 2023. *Facts and figures on the European Union.* European Union. Available at: https://european-union.europa.eu/principles-countries-history/facts-and-figures-european-union_en. (Accessed: 20 Aug 2025

Extractive Industries Transparency Initiative. 2023. Engaging communities in a just transition. Available at: https://eiti.org/sites/default/files/2023-10/EITI_Report_Engaging%20communities%20in%20a%20just%20transition.pdf. (Accessed: 6 November 2025)

Falcone, P. M. 2023. Sustainable energy policies in developing countries: a review of challenges and opportunities. *Energies*, *16*(18). https://doi.org/10.3390/en16186682

Fashion Revolution. 2023. Fashion Transparency Index. Available at: https://www.fashionrevolution.org/fashion-transparency-index-2023/. (Accessed: 20 August 2025)

Fernandes, S. 2024. Just Means "Just" Everywhere: How Extractivism Stands in the Way of an Internationalist Paradigm for Just Transitions. *International Journal of Politics, Culture, and Society*, 37 (4): 493–511. https://doi.org/10.1007/s10767-024-09475-4

Fletcher, K. and Tham, M. 2019. Earth Logic: Fashion Action Research Plan. Available at: https://katefletcher.com/wp-content/uploads/2019/10/Earth-Logic-plan-FINAL.pdf (Accessed: 20 August 2025)

Fricker, M. (2007). *Epistemic Injustice: Power and the Ethics of Knowing*. Oxford Academic. https://doi.org/10.1093/acprof:oso/9780198237907.001.0001.

Fúnez-Flores, J. *et al.* (2022). Decolonial Discourses and Practices: Geopolitical Contexts, Intellectual Genealogies, and Situated Pedagogies. *Educational Studies* 58 (5-6): 596–619. https://doi.org/10.1080/00131946.2022.2132393

General Secretariat of the Council. 2024. Draft proposal for a list of Global Gateway flagship projects for 2025 and for amending the lists of Global Gateway flagship projects for 2023 and 2024. Available at: https://data.consilium.europa.eu/doc/document/ST-15281-2024-INIT/en/pdf. (Accessed:13 August 2025);

Gireesan. K. 2024. Sustainable Governance. In: Democracy, Leadership and Governance – Application of Artificial Intelligence. Palgrave Macmillan, Singapore. https://doi.org/10.1007/978-981-99-7735-2_6.

Farrington, F. 2021. Municipio de Mariana v BHP Group: Implications of the UK High Court's Decision. *Business and H*uman Rights *Journal,* 6(2)*:* 392–8. doi:10.1017/bhj.2021.10

Fasting, S. et al. 2021. Climate Governance and Agriculture in Southeast Asia: Learning From a Polycentric Approach. *Frontiers in Political Science*, 3. https://doi.org/10.3389/fpos.2021.698431

Feguš, V. 2023. Chapter 4: EU soft law: Validity, normativity and 'bindingness' reviewed. In The Legal Effects of EU Soft Law. Cheltenham, UK: Edward Elgar Publishing.

Ferguson, J. 1994. *Anti-politics machine: Development, depoliticization, and bureaucratic power in Lesotho. University of Minnesota Press.*

Fernandes, S. 2024. Just Means "Just" Everywhere: How Extractivism Stands in the Way of an Internationalist Paradigm for Just Transitions. *International Journal of Politics, Culture, and Society*, 37 (4): 493–511. https://doi.org/10.1007/s10767-024-09475-4

Fisher, E. et al. 2019. What Is Environmental Law? In: Fisher, E., Lange, B. and Scotford, E. (Eds.) Environmental Law. Oxford University Press. 19.

Fletcher, K. and Tham, M. 2019. Earth Logic: Fashion Action Research Plan. Available at: https://katefletcher.com/wp-content/uploads/2019/10/Earth-Logic-plan-FINAL.pdf (Accessed: 20 August 2025)

Fletcher, K. 2010. Slow Fashion: An Invitation for Systems Change. *Fashion Practice* 2(2): 259–65. https://doi.org/10.2752/175693810X12774625387594

Forns Gómez, A., & Rehage, R. 2024. Is Public Participation in Hydrogen Matters (Un)just? Evaluating the Ex-/inclusion of the Vulnerable and Marginalised Individuals in the Hydrogen Decision-Chains of the EU, Spain and Germany. *Journal for European Environmental & Planning Law*, 21(3-4), 318–42. https://doi.org/10.1163/18760104-21030006

Forsyth, M. et al. 2022. Environmental Restorative Justice: An Introduction and an Invitation. In: Pali, B., Forsyth, M. and Tepper F. (eds). The Palgrave Handbook of Environmental Restorative Justice. Springer International Publishing. 3.

Fraser, N. and Honneth, A. 2003. *Redistribution or recognition?: a political-philosophical exchange*. verso.

Freedom House. 2024. Tunisia. Available at: https://freedomhouse.org/country/tunisia. (Accessed: 20 October 2025)

Gallegos, F. 2024. Bridging the Gap: Certification Schemes for Sustainable Hydrogen in Global Trade. *Journal for European Environmental & Planning Law*, 21 (3-4), 239–73. https://doi.org/10.1163/18760104-21030005

Garcia, A. S. et al. 2024. Tariff flags and electricity consumption response in Brazil. *Utilities Policy*, 88. https://doi.org/10.1016/j.jup.2024.101751.

Garment Export House. RMG Meaning in the Garment Industry. Available at: https://www.garmentexporthouse.com/2021/09/rmg-meaning-in-garment-industry.html. (Accessed: 20 August 2025)

Garetto, R. 2021. Overcoming energy poverty through becoming a prosumer? In L. Ruggeri (Ed.), Needs and Barriers of Prosumerism in the Energy Transition Era (pp. 43–55). Dykinson.

Gas connect Austria. Available at: https://www.gasconnect.at/en/about-us/about-us/our-owners. (Accessed: 20 October 2025);

Gaster, R. 2025. Small Modular Reactors: A Realist Approach to the Future of Nuclear Power. *Center for Clean Energy Innovation, Information Technology and Innovation Foundation (ITIF)*. Waschington DC: ITIF. Available at: https://itif.org/publications/2025/04/14/small-modular-reactors-a-realist-approach-to-the-future-of-nuclear-power/. (Accessed: 8 August 2025)

GFA & ALCOR. National Strategy for the Development of Green Hydrogen and its Derivatives in Tunisia. July 2024. Available at: https://www.energiemines.gov.tn/fileadmin/docs-u1/Re%CC%81sume%CC%81_strate%CC%81gie_nationale_MIME_Anglais.pdf. (Accessed: 20 October 2025)

Giobbi, M. 2021. Il consumatore energetico nel prisma del nuovo quadro regolatorio italo-eurounitario (pp. 136–7). Napoli: Edizioni Scientifiche Italiane.

Green Policy Platform. 2019. Principles, priorities and pathways for inclusive green economies: Economic transformation to deliver the SDGs. Available at: https://www.greenpolicyplatform.org/research/principles-priorities-and-pathways-inclusive-green-economies-economic-transformation. (Accessed: 20 October 2025)

Goldberg, S.M., & Rosner, R. 2011. Nuclear Reactors: Generation to Generation. *American Academy of Arts and Sciences*. Available at: https://www.amacad.org/sites/default/files/academy/pdfs/nuclearReactors.pdf. (Accessed: 11 August 2025)

Goldthau, A. and Sovacool, B. K. 2012. The Uniqueness of the Energy Security, Justice, and Governance Problem. *Energy Policy*, 41: 232–40. https://doi.org/10.1016/j.enpol.2011.10.042

Gomez, C. *et al.* 2025. What is a 'Just Transition', and are countries really making progress? World Resources Institute. Available at: https://www.wri.org/insights/what-is-just-transition-tracking-progress. (Accessed: 20 October 2025)

Gonzalez, C. 2015. Environmental Justice, Human Rights, and the Global South. *Santa Clara Journal of International Law*, 13 (1). https://digitalcommons.law.scu.edu/scujil/vol13/iss1/8.

Goodwin Brown, E. and Glutting, J. 2024. It's time to get specific on how circular economy strategies can service just transitions. Materia Rinnovabile. Available at: https://www.renewablematter.eu/en/Its-time-to-get-specific-on-how-circular-economy-strategies-can-service-just-transitions. (Accessed: 6 November 2025)

Götzmann, N. and Dicalou, M. (2025). Towards a Feminist Energy Justice Framework. *Business and H*uman Rights *Journal, 10(1): 56–78.*doi:10.1017/bhj.2025.13;

Government of Brazil. Ministry of Mines and Energy. *About Tariff Flags*. 2025. Available at: https://www.gov.br/aneel/pt-br/assuntos/tarifas/bandeiras-tarifarias. (Accessed: 14 Aug 2025)

Greenpeace Italy et. Al. v. ENI S.p.A., the Italian Ministry of Economy and Finance

Grossmann, K., and Trubina, E. 2021. How the Concept of Dignity Is Relevant to the Study of Energy Poverty and Energy Justice. *Frontiers in Sustainable Cities,* 3, 1–11. https://doi.org/10.3389/frsc.2021.644231

Gudynas, E. 2019. Value, Growth, Development: South American Lessons for a New Ecopolitics. *Capitalism Nature Socialism.* https://doi.org/10.1080/10455752.2017.1372502

Gui, E.M., MacGill, I. 2018. Typology of future clean energy communities: An exploratory structure, opportunities, and challenges. *Energy Research & Social Science,* 35: 94–107. https://doi.org/10.1016/j.erss.2017.10.019.

Guinebault, M. 2023. Rana Plaza: ten years later, has the fashion industry learned its lesson? *Fashion Network.* Available at https://ww.fashionnetwork.com/news/Rana-plaza-ten-years-later-has-the-fashion-industry-learned-its-lesson-,1506162.html (Accessed: 20 August 2025)

Hamouchene, H. and Sandwell, K. (eds). 2023. Dismantling Green Colonialism: Energy and Climate Justice in the Arab Region. Pluto Press.

Hamouchene, H. 2023. The Energy Transition in North Africa: Neocolonialism Again! In: Hamouchene H. & Sandwell, K. (Eds.), *Dismantling Green Colonialism:* Energy and Climate Justice in the Arab Region (1st ed., pp. 29–48). Pluto Press. Available at: https://doi.org/10.2307/jj.7583922.7

Hamouchene, H. *et al.* 2022. Renewable energy in Tunisia: An unjust transition. In: Hamouchene H. & Sandwell, K. (Eds.), Dismantling Green Colonialism: Energy and Climate Justice in the Arab Region (1st ed., pp. 29–48). Pluto Press. Available at: https://doi.org/10.2307/jj.7583922.15

Hazrati, M. and Annamayer, E., 2025. Evaluating EU Directives on Corporate Sustainability Reporting and Due Diligence Through the Lens of the 'JUST' Framework. In New Engagement Strategies for Energy Justice: Perspectives from the Next Generation (pp. 121–41). Cham: Springer Nature Switzerland.

Handrlica, J., & Blahoudková, G. 2025. Microreactors for powering data centres in smart cities, miniaturisation of technologies and the paradigm shift in nuclear law. Available at: https://ceridap.eu/microreactors-for-powering-data-centres-in-smart-cities-miniaturisation-of-technologies-and-the-paradigm-shift-in-nuclear-law/ (Accessed: 29 June 2025).

Haneef, M.A.(2003). Decolonizing Methodologies: Research and Indigenous Peoples by Linda Tuhiwai Smith (London, UK: Zed Books Ltd/ University of Otago Press, 1999: 208 pages. *American Journal of Islam and Society,* 20 (3-4): 188–91. https://doi.org/10.35632/ajis.v20i3-4.1836

Hanke, F., and Guyet, R. 2023. The Struggle of Energy Communities to Enhance Energy Justice: Insights from 113 German cases. *Energy Sustainable Society* 13, 16. https://doi.org/10.1186/s13705-023-00388-2

Hanke, F., Guyet, R., & Feenstra, M. 2021. Do renewable energy communities deliver energy justice? Exploring insights from 71 European cases. *Energy Research & Social Science, 80,* 102244.

Haque, C. E. *et al.* 2019. "Events and failures are our only means for making policy changes": learning in disaster and emergency management policies in Manitoba, Canada. *Natural Hazards*, 98(1): 137–62. https://doi.org/10.1007/s11069-018-3485-7

Haraway, D. (1988). Situated Knowledges: The Science Question in Feminism and the Privilege of Partial Perspective. *Feminist Studies*, 14 (3): 575–99. https://doi.org/10.2307/3178066

Hargreaves, S. *et al.* 2025. A Critical Exploration of Pan-African Ecofeminist Popular Education within WoMin's Feminist Schools. *Studies in the Education of Adults*, 1–20. https://doi.org/10.1080/02660830.2024.2445905

Hazrati, M. and Heffron, R.J. (2021). Conceptualising Restorative Justice in the Energy Transition: Changing the Perspectives of Fossil Fuels. *Energy Research and Social Science*, 78. https://doi.org/10.1016/j.erss.2021.102115

Healy, N., *et al.* 2019. Embodied Energy Injustices: Unveiling and politicizing the transboundary harms of fossil fuel extractivism and fossil fuel supply chains. *Energy Research & Social Science* 48: 219–34. https://doi.org/10.1016/j.erss.2018.09.016

Healey, P. 2010. Making better places: The planning project in the twenty-first century. Palgrave Macmillan; European Commission. 2016. Urban Agenda for the EU: Pact of Amsterdam. Brussels: European Union.

Heffron, R. J., and Hazrati, M. 2024. The Rise of Energy Justice Research in the Energy Transition and for Climate Mitigation. Just Transitions. Palgrave Macmillan, Cham. https://doi.org/10.1007/978-3-031-57304-0_1

Heffron, R. 2015. Deconstructing Energy Law and Policy: The case of nuclear energy. Edinburgh: Edinburgh University Press.

Heffron, R.J., and McCauley, D. 2014. Achieving Sustainable Supply Chains Through Energy Justice. *Applied Energy* 123: 435–7. https://doi.org/10.1016/j.apenergy.2013.12.034.

Heffron, R. J. 2022. Applying Energy Justice into the Energy Transition. *Renewable and Sustainable Energy Reviews*, 156. https://doi.org/10.1016/j.rser.2021.111936

Heffron, R.J. & de Fontenelle, L. (eds.) 2024. The Power of Energy Justice & the Social Contract. *Palgrave Macmillan*. https://doi.org/10.1007/978-3-031-46282-5;

Heffron, R.J. (2023). Energy Justice–the Triumvirate of Tenets Revisited and Revised. *Journal of Energy & Natural Resources Law*, 42(2): 227–33. https://doi.org/10.1080/02646811.2023.2256593

Heffron, R.J. and McCauley, D. 2017. The concept of energy justice across the disciplines. *Energy policy*, *105*, pp.658–67.

Heffron, R. J. and McCauley, D. (2018). What Is the 'Just Transition'? *Geoforum*, 88:74–7. https://doi.org/10.1016/j.geoforum.2017.11.016

Heffron, R. J., McCauley, D., & Sovacool, B. K. 2015. Resolving society's energy trilemma through the Energy Justice Metric. *Energy Policy*, 87, 168–76.

Heffron, R. J. 2015. Deconstructing Energy Law and Policy. Edinburgh: Edinburgh University Press.

Heffron, R.J., and McCauley, D. 2014. Achieving sustainable supply chains through energy justice. *Applied Energy*, 123, 435–7. https://doi.org/10.1016/j.apenergy.2013.12.034

Heffron, R. J., and De Fontenelle, L. 2023. Implementing Energy Justice Through a New Social Contract. *Journal of Energy & Natural Resources Law*, 41 (2): 141–55. https://doi.org/10.1080/02646811.2023.2186626.

Heffron, R .J. and Sokolowski, M. M. 2024. Resolving Energy Policy Failure: Introducing energy justice as the solution to achieve a just transition. *Energy Policy*,187. https://doi.org/10.1016/j.enpol.2024.114042;

Heffron, R. 2015. Deconstructing Energy Law and Policy: The case of nuclear energy. Edinburgh: Edinburgh University Press.

Heffron, R. J. *et al.* 2023. Pathways of Scholarship for Energy Justice and the Social Contract. *Journal of Energy & Natural Resources Law*: 41 (2): 211–32. https://doi.org/10.1080/02646811.2023.2190689.

Heffron, R., and Little, G.F.M. 2016. Delivering Energy Law and Policy in the EU and US: A Reader. Edinburgh: Edinburgh University Press.

Heffron, R.J. 2024. Energy Justice - The First Step in an Energy Decision Today. In: Heffron, R.J. and de Fontenelle, L. (eds). The Power of Energy Justice & the Social Contract. Cham: Palgrave Macmillian, 7. https://doi.org/10.1007/978-3-031-46282-5_1

Heffron, R. 2021. Achieving a Just Transition to a Low-Carbon Economy. Palgrave Macmillan Cham

Heffron, R. J. and Talus, K. 2016. The Evolution of Energy Law and Energy Jurisprudence: Insights for Energy Analysts and Researchers. *Energy Research & Social Science*, 19: 1–10. https://doi.org/10.1016/j.erss.2016.05.004;

Heiko, K. 2016. European Energy Law and Policy: An Introduction. Cheltenham, UK: Edward Elgar Publishing;

Hemingway, E. 1932. Death in the Afternoon (Scribner's 1932) 192–3 (explaining the "Iceberg Theory")

Hertwich, E. G. 2005. Consumption and the Rebound Effect: An Industrial Ecology Perspective. *Journal of Industrial Ecology* 9(1–2): 85–98. https://doi.org/10.1162/1088198054084635

Hickel, J. 2021. Less Is More: How Degrowth Will Save the World. London: William Heinemann

Hickel, Jason. 2021. Less Is More: How Degrowth Will Save the World. London: William Heinemann

Hickel, J. 2021. What does degrowth mean? A few points of clarification. *Globalizations* 18(7): 1105–11. https://doi.org/10.1080/14747731.2020.1812222

Hickel, J. and Kallis, G. 2020. Is Green Growth Possible? *New Political Economy* 25(4): 469–86. https://doi.org/10.1080/13563467.2019.1598964

Hivos. 2025. Breakthrough for energy democracy in Tunisia. Available at: https://hivos.org/breakthrough-for-energy-democracy-in-tunisia/.(Accessed: 20 October 2025)

Hivos. An insider's view on just climate action in Tunisia. Available at: https://hivos.org/story/an-insiders-view-on-just-climate-action-in-tunisia/. (Accessed: 20 October 2025)

Hochstetler, R. L., & Born, P. H. S. 2022. Community energy design models in Brazil: From niches to mainstream. In:S. Löbbe, F. Sioshansi, & D. Robinson (Eds.), Energy Communities: Customer-Centered, Market-Driven, Welfare-Enhancing? (pp. 317–38). Academic Press.

Hoesch, K. *et al.* (2025). What to Expect When You're Expecting Engagement: Delivering Procedural Justice in Large-Scale Solar Energy Deployment. *Energy Research & Social Science*, 120. https://doi.org/10.1016/j.erss.2024.103893

Höffken, J. and Ramana, M.V. 2024. Nuclear Power and Environmental Injustice. *Wiley Interdisciplinary Reviews: Energy and Environment* 13(1): https://doi.org/10.1002/wene.498.

Hoffmann, M.M., Kammen, D.M., Blechinger, P. 2025. Are Energy-Poor Communities in California More Vulnerable to Heat Wave-Related Health Burdens?, in: Phillips, A., Kaschny, L. (Eds.), New Engagement Strategies for Energy Justice: Perspectives from the Next Generation. *Springer Nature Switzerland 2025*

Holman, R. and Lefort, J. 2024. Invested in Renewables: The only way forward. European Investment Bank. Available at: https://www.eib.org/en/essays/europe-energy-transition-renewable. (Accessed: 6 November 2025)

Ho, J. et al. 2014. Perceived environmental and health risks of nuclear energy in Taiwan after Fukushima nuclear disaster. *Environmental International*, 73: 295–303. https://doi.org/10.1016/j.envint.2014.08.007.

Huhta, K. 2025. An Essential Zero-Carbon Solution or a Perilous Energy Technology? Exploring the nuclear debate in EU law. In: Reins, L. and Zahar, A. (eds) Climate Technology and Law in the Anthropocene. Bristol: Bristol University Press.

Human Rights Watch. 2021. Tunisia: President's Seizure of Powers Threatens Rights: A dangerous move capitalizes on popular frustration. Available at: https://www.hrw.org/news/2021/07/27/tunisia-presidents-seizure-powers-threatens-rights. (Accessed: 20 October 2025)

Hydrogen Europe. 2024. Six hydrogen production deals concluded in Tunisia. Available at: https://hydrogeneurope.eu/six-hydrogen-production-deals-concluded-in-tunisia/ (Accessed: 12 May 2025).

IAEA. 2022. Small Modular Reactors: A new nuclear energy paradigm. Vienna: IAEA. Available at: http://nucleus.iaea.org/sites/smr/Shared%20Documents/Small%20Modular%20Reactors%20a%20new%20nuclear%20energy%20paradigm.pdf. (Accessed: 8 May 2025).

IAEA. 2021. Nuclear Energy for a Net-Zero World. Vienna: IAEA. Available at: https://www.iaea.org/sites/default/files/21/10/nuclear-energy-for-a-net-zero-world.pdf. (Accessed: 11 May 2025).

IAEA. 2021. Nuclear Energy for a Net-Zero World. Vienna: IAEA. Available at: https://www.iaea.org/sites/default/files/21/10/nuclear-energy-for-a-net-zero-world.pdf. (Accessed: 11 May 2025).

IAEA. 2023. Decommissioning of Nuclear Installations. Vienna: IAEA. Available at: https://www.iaea.org/topics/decommissioning. (Accessed: 11 August 2025).

IBGE – News Agency. (2023). 2022 Census: 87% of the Brazilian population lives in urban areas. IBGE News Agency. Available at: https://agenciadenoticias.ibge.gov.br/en/agencia-news/2184-news-agency/news/41910-2022-census-87-of-the-brazilian-population-lives-in-urban-areas. (Accessed: 30 Jul 2025).

Ibrahima, A.B. and Mattaini, M.A. (2018). Social Work in Africa: Decolonizing Methodologies and Approaches. *International Social Work,* 62 (2): 799–813. https://doi.org/10.1177/0020872817742702

Iglesias, C., & Vilaça, P. 2022. On the regulation of solar distributed generation in Brazil: A look at both sides. *Energy Policy,* 167. https://doi.org/10.1016/j.enpol.2022.113091.

IHRB. 2023. Rana Plaza 10 Years On: Lessons for Human Rights and Business. Available at: https://ihrb-org.files.svdcdn.com/production/assets/uploads/briefings/rana-plaza-lessons-for-human-rights-and-business.pdf. (Accessed: 20 August 2025)

ILO. 2023. The Rana Plaza disaster ten years on: What has changed? Available at: https://webapps.ilo.org/infostories/en-GB/Stories/Country-Focus/rana-plaza#intro (Accessed: 20 August 2025)

ILO. n.d. A turning point for occupational safety and health in supply chains. Available at: https://vzf.ilo.org/Photo%20Exhibition/a-turning-point-for-occupational-safety-and-health-in-supply-chains/#:~:text=After%20the%20disaster%2C%20the%20global,of%20profit%20and%20economic%20growth. (Accessed: 20 August 2025)

ILO. n.d. The Sustainability Compact for the Bangladesh Ready-Made Garment Sector. Available at: https://mhssn.igc.org/EU%20Sustain%20Compact%20-%20BGD%20-%20Jan%202019%20.pdf (Accessed: 20 August 2025)

industriAll Europe. 2023. Rana Plaza tragedy: a turning point for a safer garment industry. Available at: https://news.industriall-europe.eu/Article/893. (Accessed: 20 June 2025)

Integrated Energy, Climate and Spatial planning to enable local and regional authorities to effectively implement their plan. 2.1. Renewable energy. Available at: https://ieecp.org/inplanpractice/focus-topics-for-climate-action/renewable-energy/. (Accessed: 7 November 2025)

International Commission of Jurists. 2024. Tunisia: Unfair Presidential Election Undermines Human Rights and Deepens Political Crisis. Available at: https://

www.icj.org/tunisia-unfair-presidential-election-undermines-human-rights-and-deepens-political-crisis/ Accessed: 27 July 2025)

Inter-American Court of Human Rights. Case of the Inhabitants of La Oroya v Peru. Available at: https://www.corteidh.or.cr/docs/casos/articulos/seriec_511_ing.pdf. (Accessed: 7 November 2025)

International Atomic Energy Agency (IAEA). 2023. What are Small Modular Reactors (SMRs)? Vienna: IAEA. Available at: https://www.iaea.org/newscenter/news/what-are-small-modular-reactors-smrs. (Accessed: 6 May 2025)

International Energy Agency. Russia's war on Ukraine: Analysing the impacts on energy markets and energy security. Available at: https://www.iea.org/topics/russias-war-on-ukraine. (Accessed: 6 November 2025)

International Finance Corporation. 2019. Tackling Childcare: The Business Benefits and Challenges of Employer-Supported Childcare in Bangladesh. Available at: https://www.ifc.org/content/dam/ifc/doc/mgrt/final-ifc-tackling-childcare-bangladesh-110619.pdf. (Accessed: 20 August 2025).

IPCC. 2022. Climate Change 2022: Mitigation of Climate Change. Available at: https://www.ipcc.ch/report/ar6/wg3/downloads/report/IPCC_AR6_WGIII_FullReport.pdf (Accessed: 20 August 2025)

Islam, M. 2023. The Importance of Research in the Advancement of Knowledge and Society. International Research Journal of Basic and Clinical Studies, 8(4): 1–3. DOI: 10.14303/irjbcs.2023.49

Istituto Leonarda Vaccari. 2025. Available at: https://www.leonardavaccari.it. (Accessed:31 July 2025).

Italy. 2017. Legislative Decree No. 117 of 3 July 2017, Code of the Third Sector, Article 55.

Italy. 2023. Legislative Decree No. 36 of 31 March 2023, Public Contracts Code, Article 2 (Principle of Trust).

Izam, N. S. M. N., *et al.* 2022. Sustainable development perspectives of solar energy technologies with focus on solar Photovoltaic—A review. *Energies*, *15*(8). https://doi.org/10.3390/en15082790.

Jackson, T. 2016. Prosperity Without Growth: Foundations for the Economy of Tomorrow. Oxford: Taylor & Francis Group

Jenkins, K. E. H. *et al.* 2021. The Methodologies, Geographies, and Technologies of Energy Justice: A Systematic and Comprehensive Review. *E*nvironmental *Research Letters*, 16 (4). DOI 10.1088/1748–326/abd78c;

Jenkins, K., McCauley, D., Heffron, R., Stephan, H., Rehner, R. 2016. Energy justice: A conceptual review. *Energy Res. Soc. Sci.*, 11, 174–82. https://doi.org/10.1016/j.erss.2015.10.004;

Jenkins, K., Sovacool, B. K., & McCauley, D. 2018. Humanizing sociotechnical transitions through energy justice: An ethical framework for global transformative change. *Energy policy*, *117*, 66–74

Jiménez Casanova, P. 2024. Towards a Sustainable, Integrated, and Decarbonized Energy System in the EU: Addressing Structural Challenges Through

Hydrogen System Planning. *Journal for European Environmental & Planning Law, 21*(3-4), 193–214. https://doi.org/10.1163/18760104-21030004.

Judah, B., *et al.* 2023. How to Strike a Grand Bargain on EU Nuclear Energy Policy. *Atlantic Council.* Available at: https://www.atlanticcouncil.org/in-depth-research-reports/report/how-to-strike-a-grand-bargain-on-eu-nuclear-energy-policy/. (Accessed:15 May 2025)

Juskus, R. 2023. Sacrifice Zones: A Genealogy and analysis of an environmental justce concept. *Environmental Humanities*, 15(1): 3–24. https://doi.org/10.1215/22011919-10216129

Kaschny, L. 2023. Energy Justice and the Principles of Article 194(1) TFEU Governing EU Energy Policy. *Transnational Environmental Law*: 23(2): 270–94. doi:10.1017/S2047102523000110

Kaschny, L., Phillips, A. Lessons from New Engagement Strategies for Energy Justice, in: Phillips, A., Kaschny, L. (Eds.), New Engagement Strategies for Energy Justice: Perspectives from the Next Generation. *Springer Nature Switzerland*

Kamphuis, C. 2012. Foreign Mining, Law and the Privatization of Property: A Case Study from Peru. *Journal of Human Rights and the Environment*, 3(2): 217, 222.

Katsiotis, N. 2025. Small Modular Reactors: Redefining global energy security for a digitalized future. *Strategy International.* Nicosia: Cyprus. Available at: https://strategyinternational.org/2025/02/05/publication161/. (Accessed: 17 May 2025).

Kelleher, Dylan. 2025. Fast Fashion. Brittanica, 2025. Available at: https://www.britannica.com/art/fast-fashion. (Accessed: 20 August 2025)

Khaleel, M., et al. 2024. Impact of Smart Grid Technologies on Sustainable Urban Development. *International Journal of Electrical Engineering and Sustainability (IJEES)*, 2(2), 62–82. DOI: 10.5281/zenodo.11577746

Khan, F. and Heinecker, P. 2018. Inequality and energy: Revisiting the relationship between disparity of income distribution and energy use from a complex systems perspective. *Energy Research & Social Science, 42*, pp.184–92

Khan, T. H., & MacEachen, E. 2021. Foucauldian Discourse Analysis: Moving Beyond a Social Constructionist Analytic. *International Journal of Qualitative Methods, 20*, 16094069211018009. https://doi.org/10.1177/16094069211018009

Kirchherr, J. et al. 2023. Conceptualising the Circular Economy (Revisited): An Analysis of 221 definitions. *Resources, Conservation & Recycling*, 194. https://doi.org/10.1016/j.resconrec.2023.107001.

Koukoufikis, G., *et al.* 2023. Energy Communities and Energy Poverty. Publications Office of the European Union, Luxembourg. doi:10.2760/389514, JRC134832

Krall, L. M., *et al.* 2022. Nuclear Waste from Small Modular Reactors. *Proceedings of the National Academy of Sciences:* 119 (23). https://doi.org/10.1073/pnas.2111833119.

Kumar, K., Höffken, J. and Pols, A. (Eds.). (2021). *Dilemmas of Energy Transitions in the Global South: Balancing Urgency and Justice* (1st ed.) Routledge. https://doi.org/10.4324/9780367486457;

Lacey-Barnacle, M. *et al.* (2020). Energy Justice in the Developing World: A Review of Theoretical Frameworks, Key Research Themes and Policy Implications. *Energy for Sustainable Development*, 55: 122–38. https://doi.org/10.1016/j.esd.2020.01.010

Lakic, I. et al. 2023. A just energy transition takes a ecosystem. Boston Consulting Group. Available at: https://www.bcg.com/publications/2023/exploring-challenges-of-just-transition. (Accessed: 20 October 2025)

Latouche, S. 2009. Farewell to growth. Cambridge: Polity Press; Vanderheiden, Steve. 2020. *Environmental Political Theory*. Cambridge: Polity Press

Leavy, P. 2014. Introduction to the Oxford Handbook of Qualitative Research, Second Edition. In: Leavy, P (ed.) The Oxford Handbook of Qualitative Research. Oxford University Press.

Le Gouais, A. and Cooper, M. 2024. Who can influence healthy urban development? *Cities & Health*, 9 (4): 644–56. https://doi.org/10.1080/23748834.2024.2362508

Leijon, K. 2020. It Takes Two to Tango: The preliminary reference dance between the Court of Justice of the European Union and national courts, ed. J. Krommendijk. *European Papers* 5 (2): 877–902. Available at: https://www.europeanpapers.eu/en/system/files/pdf_version/EP_eJ_2020_2_10_Articles_SS1_7_Karin_Leijon_00416.pdf. (Accessed; 11 August 2025).

Levine, T. Quantitative Social Science Methods of Inquiry. In: Knapp, M.L. and Daly, J.A. (eds). The Sage Handbook of Interpersonal Communication. 4th Edition. Sage Publications

Li, F. 2015. Unearthing Conflict: Corporate Mining, Activism, and Expertise in Peru. Duke University Press, 15–20.

Li, Tania Murray. 2017. After development: Surplus population and the politics of entitlement. *Development and Change* 48.6, 1247–61.

Lindt, A. 2020. Transnational Human Rights Litigation: A Means of Obtaining Effective Remedy Abroad?. *Journal of Legal Anthropology*. https://doi.org/10.3167/jla.2020.040204

Lifset, R. et al. 2023. Restoring the incentives for eco-design in extended producer responsibility: The challenges for eco-modulation. *Waste Management* 168: 189–201. https://doi.org/10.1016/j.wasman.2023.05.033

Listo, R. (2018). Gender Myths in Energy Poverty Literature: A Critical Discourse Analysis. *Energy Research and Social Science, 38:* 9–18. https://doi.org/10.1016/j.erss.2018.01.010

Liu, K. 2024. Circular economy and the separated yet inseparable social dimension: Views from European circular city experts. *Sustainable Production and Consumption*, 51: 474–83. https://doi.org/10.1016/j.spc.2024.09.016.

Löfquist, L. 2020. Is there a universal human right to electricity? *The International Journal of Human Rights*, 24(6): 711–23. https://doi.org/10.1080/13642987.2019.1671355

Lohmeyer, N. and Schüßler, E. 2019. Rana Plaza as a Threat to the Fast Fashion Model? An Analysis of Institutional Responses to the Disaster in Germany. In: Becker-Leifhold, and M. Heuer (Eds.) Eco Friendly and Fair: Fast Fashion and Consumer Behavior. 3–14. London and New York: Routledge

Lotter, A. 2020. Part 2: Mine(d) Over Matter: Restorative Justice in the Mining Industry. Available at: https://law.uct.ac.za/mineral-law/articles/2020-02-05-part-2-mined-over-matter-restorative-justice-mining-industry (Accessed: 25 June 2025)

Longeran, K., *et al.* 2022. Energy Justice Analysis of the European Commission REPowerEU plan. *Working Paper*. Zurich: ETH Zurich. https://doi.org/10.3929/ethz-b-000551952

Mahajan, R., & Bandyopadhyay, K. R. 2021. Women entrepreneurship and sustainable development: select case studies from the sustainable energy sector. *Journal of Enterprising Communities: People and Places in the Global Economy*, *15*(1), 42–75. DOI: 10.1108/JEC-11-2020-0184

MacArthur, J. 2016. Challenging public engagement: Participation, deliberation and power in renewable energy policy. *Journal of Environmental Studies and Sciences*, 6(3), 631–40.

Malin, S. A., & Ryder, S. S. 2018. Developing deeply intersectional environmental justice scholarship. *Environmental Sociology*, *4*(1), 1–7.

Mallick, P. K. et al. 2024. Designing and operationalising extended producer responsibility under the EU Green Deal. *Environmental Challenges*, 16 https://doi.org/10.1016/j.envc.2024.100977

Martinelli, C. 2025. A just future needs gender and climate justice. Climate Action Network. Available at: https://caneurope.org/a-just-future-needs-gender-and-climate-justice/. (Accessed: 6 November 2025)

Marti, E. et al. 2024. The Impact of Sustainable Investing: A Multidisciplinary Review. *Journal of Management Studies* 61(5): 2181–211. https://doi.org/10.1111/joms.12957

Martin Morán, A. 2024. Diversifying Clean Hydrogen. Regulatory Barriers for a Circular Hydrogen Economy. *Journal for European Environmental & Planning* Law, 21 (3-4), 167 – 92. https://doi.org/10.1163/18760104-21030003.

Martin, P. 2024. Tunisia's new national green hydrogen strategy sets out plans for first H2 exports to Europe by 2030. Hydrogen Insight. Available at: https://www.hydrogeninsight.com/policy/tunisias-new-national-green-hydrogen-strategy-sets-out-plans-for-first-h2-exports-to-europe-by-2030/2-1-1651045. (Accessed: 20 October 2025)

Martins, M. L. R., & Santos Pereira, A. L. D. 2019. Urban Regeneration in the Brazilian urban policy agenda. *European Planning Studies*, *27*(6), 1129–45. DOI: 10.1080/09654313.2019.1598021

Mayanti, B. and Helo, P. 2023. Circular economy through waste reverse logistics under extended producer responsibility in Finland. *Waste Management & Research* 42(1): 59–73. https://doi.org/10.1177/0734242X231168801

McCauley, D., et al. 2013. Advancing energy justice: The triumvirate of tenets. *International Energy Law Review*, 32(3), 107–16

McCauley, D. and Heffron, R. 2018. Just transition: Integrating climate, energy and environmental justice. *Energy Policy* 119: 1–7. https://doi.org/10.1016/j.enpol.2018.04.014

McCauley, D. 2018. Global energy justice: tackling systems of inequality in energy production and consumption. *Energy Justice: Re-Balancing the Trilemma of Security, Poverty and Climate Change*, pp.1–26

McCauley, D. and Heffron, R. (2018). Just Transition: Integrating Climate, Energy and Environmental Justice. *Energy Policy*, 119: 1–7. https://doi.org/10.1016/j.enpol.2018.04.014;

McCauley, D., Ramasar, V., Heffron, R. J., Sovacool, B. K., Mebratu, D., & Mundaca, L. (2019). Energy justice in the transition to low carbon energy systems: Exploring key themes in interdisciplinary research. Applied Energy, 233-234, 916–21. doi: https://doi.org/10.1016/j.apenergy.2018.10.005

McGowan, K., & Antadze, N. 2023. Recognizing the dark side of sustainability transitions. *Journal of Environmental Studies and Sciences*, *13*(2), 344–9.

McHarg, A. 2020. Energy Justice: Understanding the 'ethical turn' in energy law and policy. In: del Guayo, I. *et al.* (eds). Energy Justice and Energy Law. Oxford: Oxford University Press

Mejía-Montero, A. et al. 2021. Grounding the Energy Justice Lifecycle Framework: An Exploration of Utility-Scale Wind Power in Oaxaca, Mexico. *Energy Research & Social Science*, 75. https://doi.org/10.1016/j.erss.2021.102017

Menkel-Meadow, C. 2007. Restorative Justice: What Is It and Does It Work? *Georgetown Law Faculty Publications and Other Works*, 2

Mendes, E. (2024). Just Energy Transition Policies at the Subnational Level: A Critical and Propositional Analysis Based on the Just Energy Transition Plan of the State of Ceará [Políticas de Transição Energética Justa Em Nível Subnacional: Uma Análise Crítico-Propositiva a Partir Do Plano de Transição Energética Justa Do Estado Do Ceará, In Portuguese]. Master's Dissertation, Universidade Federal do Ceará. http://repositorio.ufc.br/handle/riufc/76486;

Metreau, E et al. 2024. World Bank country classifications by income level for 2024-2025. World Bank blogs. Available at: https://blogs.worldbank.org/en/opendata/world-bank-country-classifications-by-income-level-for-2024-2025. (Accessed: 20 October 2025)

Min. Z and Wimer, F. 2025.Transnational Governance of Soybean Land Use in South America: A Polycentric Approach. *Transnational* Environmental *Law*, 14 (1):145– 70. doi:10.1017/S2047102524000268.

Michaels, R., et al. 2021. The private side of transforming our world: UN Sustainable Development Goals 2030 and the role of private international law. (1st ed.). Intersentia studies on Private International Law.

Mies, M. and Shiva, V. 1993. *Ecofeminism*. Zed Books.

Milčiuvienė, S., Kiršienė, J., Doheijo, E., Urbonas, R. and Milčius, D. 2019. The role of renewable energy prosumers in implementing energy justice theory. *Sustainability*, *11*(19), p.5286.

Ministry of Science, Technology and Innovation; Avante Brasil (OSCIP). 2024. Work Plan – Energia Solidária Project [Project funded through a fostering agreement – Process No. 01245.000351/2024–96]. Available at: https://avantebrasil.ong.br/wp-content/uploads/2025/01/Plano_de_Trabalho_-_PRENERGIS_assinado.pdf. (Accessed:14 August 2025).

Monyei, C.G. *et al.* (2018). Examining Energy Sufficiency and Energy Mobility in the Global South through the Energy Justice Framework. *Energy Policy,* 119: 68–76. https://doi.org/10.1016/j.enpol.2018.04.026

Morton Turner, J. 2022. Charged: A History of Batteries and Lessons for a Clean Energy Future University of Washington Press

Movimento Dos Atingidos Por Barragens. 'Quem Somos.' Availale at: https://mab.org.br/quem-somos/. (Accessed: 29 June 2025).

Movimento Dos Atingidos Por Barragens. 2025. 'Mulheres e Arpilleras.' Available at: https://mab.org.br/en/women/. (Accessed: 20 June 2025).

Movimento Dos Atingidos Por Barragens. 2017. 'Mulheres Atingidas Em Defesa Da Vida.' Available at: https://mab.org.br/arpilleras-do-mab/mulheres-atingidas-em-defesa-da-vida/. (Accessed: 29 June 2025).

Movimento Dos Atingidos Por Barragens. 2019. 'Privatização Que Mata.' Available at: https://mab.org.br/arpilleras-do-mab/privatizacao-que-mata/. (Accessed: 29 June 2025).

Nakanwagi, S. 2024. Critical Minerals, Sustainability, and the Energy Transition in the Global South: A Justice Perspective. Hart. 135–6.

Natarajan, U. 2022. Who Do We Think We Are?: Human Rights in a Time of Ecological Change In: Natarajan, U. and Dehm, J. (Eds) Locating Nature: Making and Unmaking International Law. Cambridge University Press, 207.

Neij, L. *et al.* 2025. Energy communities—lessons learnt, challenges, and policy recommendations. *Oxford Open Energy*, 4. https://doi.org/10.1093/ooenergy/oiaf002.

Neumann, P. 2016. Toxic Talk and Collective (In)Action in a Company Town: The Case of La Oroya, Peru. *Social Problems*, 63(3): 431–46. https://doi.org/10.1093/socpro/spw010

Newell, P. and Mulvaney, D. 2013. The political economy of the 'just transition.' *The Geographical Journal*, 179(2): 132–40. https://doi.org/10.1111/geoj.12008

Newell, P. et al. 2023. Mind the gap: The global governance of just transitions. *Global Policy*, 14 (3): 425–37. https://doi.org/10.1111/1758-5899.13236.

Neves, C. et al. 2025. Citizens' participation in local energy communities: The role of technology as a stimulus. *European Journal of Information Systems*, *34*(1): 122–45. https://doi.org/10.1080/0960085X.2024.2302426.

North Africa Hydrogen Backbone. Sea Corridor. Available at: https://www.sea-corridor.com/en/h2-initiatives; https://www.south2corridor.net/hydrogen-production-te-h2-and-verbund-have-signed-a-mou-with-the-republic-of-tunisia/, Accessed: 13 August 2025).

Nsafon, B.E.K. *et al.* (2023) The Justice and Policy Implications of Clean Energy Transition in Africa. *Frontiers in Environmental Science,* 11. https://doi.org/10.3389/fenvs.2023.1089391

OECD. 2016. Open Government in Tunisia, OECD Public Governance Reviews, OECD, pp. 82–118. http://dx.doi.org/10.1787/9789264227118-en. (Accessed: 20 October 2025).

OECD. 2022. Proceedings of the Workshop of the Management of Spent Fuel, Radioactive waste, and Decommissioning in SMRs/ Advanced Reactor Technologies. Available at: https://www.oecd-nea.org/upload/docs/application/pdf/2024-10/4b_-_nea_rwm_r_2024_1.pdf. (Accessed: 20 October 2025).

Obrenović, S., 2025. Never-Ending Transition: Energy Justice-Development Nexus in the Western Balkans. In *New Engagement Strategies for Energy Justice: Perspectives from the Next Generation* (pp. 107–20). Cham: Springer Nature Switzerland.

Office of the High Commissioner for Human Rights. 2025. Peru: Impact of international cooperation law on NGOs. Available at: https://www.ohchr.org/en/press-releases/2025/04/peru-impact-international-cooperation-law-ngos (Accessed: 25 June 2025).

OIPE – Osservatorio Italiano sulla Povertà Energetica. 2024. La povertà energetica in Italia nel 2023. Centro Levi-Cases, University of Padova.

Olawuyi, D From Energy Consumers to Energy Citizens Legal Dimensions of Energy Citizenship. In: Fleming, R., Huhta K., and Reins, L., (eds). *Sustainable Energy Democracy and the Law,* Brill, 101–23. https://doi.org/10.1163/9789004465442_006.

Osička, J., Szulecki, K. and Jenkins, K.E. 2023. p14 Energy justice and energy democracy: Separated twins, rival concepts or just buzzwords?. *Energy Research & Social Science, 104,* p.103266.

Parkhill, K., Demski, C.C., Butler, C., Spence, A. and Pidgeon, N.F. 2013. Transforming the UK energy system: public values, attitudes and acceptability-synthesis report

Patonia, A. 2025. Green hydrogen and its unspoken challenges for energy justice. *Applied Energy,* 377, Part C. https://doi.org/10.1016/j.apenergy.2024.124674.

Pellegrini-Masini, G., Pirni, A. and Maran, S. 2020. Energy justice revisited: A critical review on the philosophical and political origins of equality. *Energy Research & Social Science, 59,* p.101310.

Perrone, N.M. 2025. Mission Impossible? Promoting Global Sustainable Development in Times of National Industrial Policies. *Global Policy.* Available at:

https://www.globalpolicyjournal.com/articles/climate-change-energy-and-sustainability/early-view-article-mission-impossible-promoting

Peterson, M. 2024. Derrida and Inheritance in Environmental Ethics: The half-lives of responsibility. Cham: Palgrave Macmillan. https://doi.org/10.1007/978-3-031-52143-0.;

Phillips, A. (2024). The Quest for Cosmopolitan Justice in the Energy Transition in Caribbean Small Island Developing States. In: Heffron, R.J. and de Fontenelle, L. (eds). The Power of Energy Justice and the Social Contract. Just Transitions. Palgrave Macmillan, Cham. https://doi.org/10.1007/978-3-031-46282-5_21;

Phillips, A., Kaschny, L. (Eds.), New Engagement Strategies for Energy Justice: Perspectives from the Next Generation. *Springer Nature Switzerland*

Piñeiro, X. *et al.* 2021. Heavy Metal Contamination in Peru: Implications on Children's Health. Nature Scientific Reports, 11, 22729. https://doi.org/10.1038/s41598-021-02163-9

Polimeni, J. M. et al. 2008. The Jevons Paradox and the Myth of Resource Efficiency Improvements. London: Earthscan

Prebisch, R. 1950. The Economic Development of Latin America and Its Principal Problems. United Nations Department of Economic Affairs, Economic Commission for Latin America (ECLA), New York.

Puig, A.P., Meyer, J.H. 2021. One Movement or Many? The Diversity of Antinuclear Movements in Europe. In: Kaiser, A., Lehtonen, M., Meyer, J.H. & Rubio, M. (eds). Engaging the Atom: The History of Nuclear Energy and Society in Europe from the 1950s to the Present, Morgantown: West Virginia University Press.

Qian, Y.*et al.* 2023. A survey on energy justice: a critical review of the literature, *Economic Research- Ekonomska Istraživanja*, 36 (3), 2155860. https://doi.org/10.1080/1331677X.2022.2155860

Quijano, A. 2000. Coloniality of power and Eurocentrism in Latin America. *International sociology, 15*(2), 215–32.

Radcliffe, S.A. (2017). Decolonising Geographical Knowledges. Transactions of the Institute of British Geographers, 42 (3): 329–33. https://doi.org/10.1111/tran.12195;

Rahman, S. and Rahman, K.M.. 2020. Multi-actor Initiatives after Rana Plaza: Factory Managers' Views. *Development and Change*, 51(5): 1331–59. https://doi.org/10.1111/dech.12572

Rainard, M., Büchs, M., Jenkins, K., & Middlemiss, L. (2025). Intersectionality in good faith: Beyond normative claims and towards practical integration in energy justice research. *Energy Research & Social Science*, 119, 103823.

Ramana, M. V., and Mian, Z. 2014. One Size Doesn't Fit All: Social priorities and technical conflicts for small modular reactors. *Energy Research & Social Science*, 2: 115–24. https://doi.org/10.1016/j.erss.2014.04.015.

Ramazan, S., Voyvoda, E., Lacey-Barnacle, M., Karababa, E., Topal, C. and Islambay, D. 2017. Energy justice - a social sciences and humanities cross-cutting theme report. *Cambridge: SHAPE ENERGY*

Razzaq I., *et al.* 2022. Reduction in energy consumption and CO2 emissions by retrofitting an existing building to a net zero energy building for the implementation of SDGs 7 and 13. *Frontiers in Environmental Science*, 10. https://doi.org/10.3389/fenvs.2022.1028793.

Regional Administrative Court, Puglia Region. (2019). *Judgment of 13 December 2019, No. 1967*

Regulation (EU) 2020/852 of the European Parliament and of the Council of 18 June 2020 on the establishment of a framework to facilitate sustainable investment and amending Regulation (EU) 2019/2088 (Taxonomy Regulation). OJL 198, 22.6.2020.

REPowerEU Plan

Reins, L., and Wijns, J. 2025. The 'Safe and Sustainable by Design' Concept - A regulatory approach for a more sustainable circular economy in the European Union? *European Journal of Risk Regulation*,16: 96–113. https://doi.org/10.1017/err.2024.29

Rich, B. 2016. The Principles of Future-proofing: A broader understanding of resiliency in the historic built environment. Available at: https://www.researchgate.net/publication/333856761_The_Principles_of_Future-Proofing_A_Broader_Understanding_of_Resiliency_in_the_Historic_Built_Environment. (Accessed: 20 October 2025)

Rinnovabili. 2025. Available at: https://www.rinnovabili.it/energia/comunita-energetiche-rinnovabili/cer-le-vele-lenergia-rinnovabile-illumina-il-futuro-sociale-di-roma/. (Accessed: 31 Jul 2025).

Robinson, C.,Simcock, N. & Petrova, S. (2023). Energy Justice and Gender. In *Handbook on Energy Justice*. Cheltenham, UK: Edward Elgar Publishing. https://doi.org/10.4337/9781839102967.00020

Rot, C. et al. 2025. Is it just planning? Unpacking markers of justice in spatial planning literature. *Journal of Planning Literature*. https://doi.org/10.1177/08854122251328259.

Ruggeri, L. 2021. Legislative Policies and Jurisprudence on Climate Change: New Tools for Removing the Barriers to New Forms of Energy Consumption. In: Ruggeri, L. (eds), 2021. Needs and Barriers of Prosumerism in the Energy Transition Era, Dickinson, 98 – 109.

Rusansky, T. 2021. Embroidering Resistance: Emotional, Embodied, Everyday Struggles of Women Affected by Dams in Brazil. *Geoforum*, 127: 126–36. https://doi.org/10.1016/j.geoforum.2021.10.006

Saadaoui, H. et al. 2024. The transition to renewable energies in Tunisia: The asymmetric impacts of technological innovation, government stability, and democracy. *Energy*,293. https://doi.org/10.1016/j.energy.2024.130686.

Sabatini, C. and Binetti, B. 2024. State Capacity, Mining and Community Relations in Peru: How Sustained Multi-Sectoral Dialogue Can Help Secure the Country's Role in Global Supply Chains. Chatam House. Available at: https://www.chathamhouse.org/sites/default/files/2024-12/2024-12-02-state-capacity-mining-community-relations-peru-sabatini-binetti.pdf. (Accessed: 20 October 2025)

Said, Z.M. and Dindar, S. 2024. Key Challenges and Strategies in the Evaluation of Sustainable Urban Regeneration Projects: Insights from a Systematic Literature Review. *Sustainability*: 16 (22). https://doi.org/10.3390/su16229903.

Sarlet, I. W. 2022. Social security in Brazil: Redefining minimum income support (No. 7). Social Law Report. Max Planck Institute for Social Law and Social Policy.

Saxena, S. B., and Baumann-Pauly, D. 2019. Off the radar: Subcontracting in Bangladesh's RMG industry. In: Sanchita B. Saxena (Eds). *Labor, Global Supply Chains, and the Garment Industry in South Asi*a, 45–61. London: Routledge

Schröder, P. 2020. Promoting a just transition to an inclusive circular economy. Chatam House. Available at: https://www.chathamhouse.org/2020/04/promoting-just-transition-inclusive-circular-economy-0/4-towards-just-circular-economy. Accessed: 22 October 2025.

Scott, M. and Powells, G. 2020. Towards a new social science research agenda for hydrogen transitions: Social practices, energy justice, and place attachment. *Energy Research & Social Science*, *61*, p.101346.

Scuto, L., 2022. The development of a new model of urban regeneration in light of the digitalization of public administration. *CERIDAP*, 2. https://doi.org/10.13130/2723-9195/2022-3-17.

Sea Corridor. North Africa Hydrogen Backbone. Available at: https://www.sea-corridor.com/en/h2-initiatives. (Accessed:13 August 2025).

Shakun, N. *et al.* (2024). Postcolonial Perspectives in the 21st Century: A Critical Analysis of Modern Theories and Methodologies. *Revista Amazonia Investiga*, 13 (78):194–208. https://doi.org/10.34069/AI/2024.78.06.16

Sharma, S., & Sengar, N. 2022. Review of solar PV training manuals and development of survey based solar PV system training formats for beginners. *Solar Energy*, *241*, 72–84. https://doi.org/10.1016/j.solener.2022.05.055.

Sharif, I.S. and Das, D. 2022. Bangladesh's Emergence as a Ready-Made Garment Export Leader: An Examination of the Competitive Advantages of the Garment Industry. *International Journal of Global Business and Competitiveness*, 17: 162–74. https://doi.org/10.1007/s42943-022-00049-9

Shepherd, L.J. (2016). Research as Gendered Intervention: Feminist Research Ethics and the Self in the Research Encounter. *Critica Contemporánea: Revista de Teoría Política*, 6: 1–15.

Silva Martinelli, F., & Lindner, A. 2021. How far are the metropolitan areas in Brazil from achieving the sustainable development goals? An analysis based on SDG dashboards. *SN Social Sciences,* 1(103). https://doi.org/10.1007/s43545-021-00131-8.

Simões, GM F., & Leder, S. M. 202. Energy poverty: The paradox between low income and increasing household energy consumption in Brazil. *Energy and Buildings, 268.* DOI:10.1016/j.enbuild.2022.112234

Skukla, P. R. *et al.* (eds.).In: *Climate Change 2022: Mitigation of Climate Change. Contribution of Working Group III to the Sixth Assessment Report of the Intergovernmental Panel on Climate Change*, ed. P.R. Shukla et. al., 1–42. Cambridge: Cambridge University Press. doi: 10.1017/9781009157926.001

Singh, D.P. and Jassi, J. Explorign the significance of statistics in the research: A comprehensive overview section A- Research paper. European Chemical Bulletin, 12: 2089–102. https://doi.org/10.31838/ecb/2023.12.s2.262

Smith, J. 2016. Imperialism in the Twenty-First Century: Globalization, Super-Exploitation, and Capitalism's Final Crisis. New York: Monthly Review Press;

Snam. Shareholders. Available at: https://www.snam.it/en/investor-relations/investire-in-snam/azionariato.html. (Accessed: 20 October 2025)

Stengers, I. 2011. Another science is possible!. *A plea for slow science*, 2011–2.

Social Electricity Tariff. 2025. Available at: https://www.gov.br/aneel/en/subjects/tariffs/social-tariff. (Accessed: 31 July 2025).

Sokołowski, M. M. and Heffron, R. J. 2022. Defining and Conceptualising Energy Policy Failure: The when, where, why, and how. *Energy Policy*, 161. https://doi.org/10.1016/j.enpol.2021.112745

SouthH_2 Corridor. Available at: https://www.south2corridor.net/ and https://www.sea-corridor.com/en/h2-initiatives. (Accessed: 13 August 2025)

Sovacool, B.K. *et al.* (2016). Energy Decisions Reframed as Justice and Ethical Concerns. *Nature Energy,* 1. https://doi.org/10.1038/nenergy.2016.24.

Sovacool, B.K. *et al.* (2017). New Frontiers and Conceptual Frameworks for Energy Justice. *Energy Policy,*105: 677–91. https://doi.org/10.1016/j.enpol.2017.03.005.

Sovacool, B.K. *et al.* (2023). Pluralizing Energy Justice: Incorporating Feminist, Anti-Racist, Indigenous, and Postcolonial Perspectives. *Energy Research & Social Science,* 97. https://doi.org/10.1016/j.erss.2023.102996;

Sovacool, B. K., and Dworkin, M.H. 2015. Energy justice: Conceptual insights and practical applications. *Applied Energy* 142: 435–44. https://doi.org/10.1016/j.apenergy.2015.01.002

Sovacool, B. K., et al.2020. Sociotechnical agendas: Reviewing future directions for energy and climate research. Energy *Research & Social Science, 70,* 101617.

Sovacool, B.K., and Ramana, M.V. 2015. Back to the Future: Small Modular Reactors, Nuclear Fantasies, and Symbolic Convergence. *Science, Technology, & Human Values*: 40 (1): 96–125. http://www.jstor.org/stable/43671221.

Sovacool, B. K. *et al.* 2019. Decarbonization and its discontents: a critical energy justice perspective on four low-carbon transitions. *Climatic Change*,155:581–619. https://doi.org/10.1007/s10584-019-02521-7

Spalding, R.J. 2023. Breaking Ground: From Extraction Booms to Mining Bans in Latin America. Oxford University Press, New York, 8.

Spillane, J. 2025. Solar Rollout in Ireland: Revolution or Reinforcing Existing Social Inequity?, in: Phillips, A., Kaschny, L. (Eds.), New Engagement Strategies for Energy Justice: Perspectives from the Next Generation. *Springer Nature Switzerland 2025*

Spinaci, S. 2025. Omnibus I - sustainability reporting - proposal amending CSRD and CSDDD - in "A new plan for Europe's sustainable prosperity and competitiveness." European Parliament Legislative Train Schedule. Available at: https://www.europarl.europa.eu/legislative-train/package-simplification-business/file-first-omnibus-package-on-sustainability-proposal-amending-csrd-and-csddd (Accessed: 13 August 2025

Statewatch. 2024. Parliamentary lawyers: democratic oversight needed for EU-Tunisia migration agreement. Available at: https://www.statewatch.org/news/2024/march/parliamentary-lawyers-democratic-oversight-needed-for-eu-tunisia-migration-agreement/ (Accessed: 20 October 2025)

Stephens, J.C. 2019. Energy democracy: redistributing power to the people through renewable transformation. *Environment: Science and Policy for Sustainable Development*, *61*(2), pp.4–13

Stevis, D. and Felli, R. 2020. Planetary just transition? How inclusive and how just? *Earth System Governance*, 6. https://doi.org/10.1016/j.esg.2020.100065

Sultana, F. 2022. Critical climate justice. *The Geographical Journal*, 188(1), 118–24.

Suraci, O. 2021. The Best-Dressed Polluter – Regulation and Sustainability in the Fashion Industry. *Hastings Environmental Law Journal* 27(2): 225

Svampa, M. 2015. Commodities Consensus: Neoextractivism and Enclosure of the Commons in Latin America. *South Atlantic Quarterly*, 114 (1): 65–7. https://doi.org/10.1215/00382876-2831290

Swedwatch. 2024. Paying the price for fashion: securing a living wage for Bangladesh's garment workers. Available at: https://swedwatch.org/wp-content/uploads/2024/11/briefsecuring-living-wagesnov-201124.pdf (Accessed: 20 August 2025)

SLYCAN Trust. 2021. Challenges and Opportunities for a Just Transition. Available at: https://weadapt.org/wp-content/uploads/2023/05/issues_paper_on_challenges_and_opportunities_for_a_just_transition_0_0.pdf. (Accessed: 6 November 2025)

Taekema, S., and van der Burg, W. 2024. Chapter 1: Introduction: Methodology of legal research. Contextualising Legal Research (Edward Elgar Publishing)

Talus, K. *et al.* 2024. Realism at the end of the rainbow? An argument towards diversifying hydrogen in EU regulation. *Journal of World Energy Law and Business,* 17 (4): 217–33. https://doi.org/10.1093/jwelb/jwae007.

Tanaka Y. 2004. Major Psychological Factors Determining Public Acceptance of the Siting of Nuclear Facilities. *Journal of Applied Social Psychology*, 34: 1147–65. https://doi.org/10.1111/j.1559-1816.2004.tb02000.x.

The Accord on Fire and Building Safety in Bangladesh. n.d. The Accord on Fire and Building Safety in Bangladesh. https://bangladeshaccord.org (Accessed:20 August 2025)

The Alliance for Bangladesh Worker Safety. n.d. The Alliance for Bangladesh Worker Safety. https://bangladeshworksafety.org (Accessed: 20 August 2025)

The Atlas of Economic Complexity. 2025. Peru. Available at: https://atlas.hks.harvard.edu/countries/604. (Accessed: 25 June 2025).

The Atlas of Economic Complexity. 2025. Copper Ore in Peru. Available at Copper Ore in Peru Trade | The Observatory of Economic Complexity. (Accessed: 8 August 2025)

The Global Gateway. Information on the roll out of the Global Gateway strategy, partnerships, projects and funding opportunities. Available at: https://international-partnerships.ec.europa.eu/policies/global-gateway_en. (Accessed: 28 May 2025).

Todorović, I. 2024. EU remains divided over nuclear energy expansion initiative. *Balkan Green Energy News*. Available at: https://balkangreenenergynews.com/eu-remains-divided-over-nuclear-energy-expansion-initiative/. (Accessed: 28 May 2025)

Tornel, C. *et al.* (2022). Decolonizing Energy Justice from the Ground up: Political Ecology, Ontology, and Energy Landscapes. *Progress in Human Geography*, 47(1). https://doi.org/10.1177/03091325221132561.

Tuler, S., and Webler, T. 2024. The Challenge of Community Acceptance of Small Modular Reactors. *Energy Research & Social Science*, 118: 1. https://doi.org/10.1016/j.erss.2024.103831.

United Nations. 2015. Transforming our world: The 2030 Agenda for Sustainable Development. New York: United Nations.

United Nations, Guiding Principles on Business and Human Rights: Implementing the United Nations "Protect, Respect and Remedy" Framework, Office of the High Commissioner for Human Rights, [ST/]HR/PUB/11/4, New York; Geneva, 2011, p. 2. Available at: https://digitallibrary.un.org/record/720245?ln=en. (Accessed: 20 October 2025)

United Nations. Department of Economic and Social Affairs. Sustainable Development Goal 11: Make cities and human settlements inclusive, safe, resilient and sustainable. Available at: https://sdgs.un.org/goals/goal11. (Accessed: 7 November 2025)

United Nations Human Rights Council .2018. UN framework principles on human rights and the environment. A/HRC/37/59

United Nations 2015. Paris Agreement. [ONLINE] Available at: https://unfccc.int/sites/default/files/english_paris_agreement.pdf (Accessed: 20 July 2025).

UN Global Compact. Rana Plaza Factory: Global Outcry Triggers Industry Action. Business & Human Rights Navigator. Available at: https://bhr-navigator.unglobalcompact.org/case_studies/global-outcry-triggers-industry-action/ (Accessed: 20 August 2025).

UN-Habitat Flagship Programme. 2024. Urban regeneration for localizing the Sustainable Development Goals. UN-Habitat. Available at: https://unhabitat.org/sites/default/files/2024/05/ur_-_paper_22052024.pdf (Accessed: 19 Aug 2025).

Vaibhav, H. 2025. Circular Economy: The Moder Hack for (Almost) Infinite Material Life. Available at: https://earth.org/circular-economy-the-modern-hack-for-almost-infinite-material-life/. (Accessed: 20 October 2025)

Valencia, A. 2014. Human Rights Trade-Offs in a Context of "Systemic Lack of Freedom": The Case of the Smelter Town of La Oroya, Peru. *Journal of Human Rights*, 13 (4). https://doi.org/10.1080/14754835.2014.960962.

Valkenburg, G. (2024). Energy Justice as Epistemic Justice. *Ethics, Policy & Environment*, 1–22. https://doi.org/10.1080/21550085.2024.2418789

Vanderheiden, S. 2020. *Environmental Political Theory*. Cambridge: Polity Press

Van Der Schoor, T., & Scholtens, B. 2015. Power to the people: Local community initiatives and the transition to sustainable energy. *Renewable and sustainable energy reviews*, *43*: 666–75. https://doi.org/10.1016/j.rser.2014.10.089.

van den Broek et al. 2025. Expert and citizen perceptions of the drivers of the energy transition: A mental model approach. *Journal of Cleaner Production*, 494, 144949

V-DEM Institute. 2025. Democracy Report 2025. 25 years of Autocratization – Democracy Trumped?,p. 33. Available at: https://www.v-dem.net/documents/61/v-dem-dr__2025_lowres_v2.pdf. (Accessed: 20 October 2025)

Van Hee, N., *et al.* 2024. Economic Potential and Barriers of Small Modular Reactors in Europe. *Renewable and Sustainable Energy Reviews*, 203. https://doi.org/10.1016/j.rser.2024.114743.

Van Uffelen, N. (2022). Revisiting Recognition in Energy Justice. *Energy Research & Social Science, 92*. https://doi.org/10.1016/j.erss.2022.102764

Van Uffelen, N. *et al.* (2024). Revisiting the Energy Justice Framework: Doing Justice to Normative Uncertainties. *Renewable and Sustainable Energy Review, 189, Part A*. https://doi.org/10.1016/j.rser.2023.113974;

van Uffelen, N. 2024. p1 Understanding energy conflicts: From epistemic disputes to competing conceptions of justice. Energy Research & Social Science, 118, p.103809.

Vazi, B. and Laan, T. 2025. Why Gender inclusion must drive the Just energy transition. International Institute for Sustainable Development. Available at: https://www.iisd.org/articles/insight/why-gender-inclusion-must-drive-just-energy-transition. (Accessed: 6 November 2025)

Veltmeyer, H. 2022. Extractivism and Beyond. The Extractive Industries and Society. https://doi.org/10.1016/j.exis.2022.101132

Velenturf, A.P.M. and Purnell, P. 2021. Principles for a sustainable circular economy. *Sustainable Production and Consumption*, 27: 1437–57. Available at: https://www.sciencedirect.com/science/article/pii/S2352550921000567. (Accessed: 22 October 2025)

Volodzkiene, L. and Streimikiene, D. 2023. Energy inequality indicators: a comprehensive review for exploring ways to reduce inequality. *Energies*, *16*(16), p.6075.

Walk, P. (2024). From Parity to Degrowth: Unpacking Narratives of a Gender-Just Transition. *Energy Research & Social Science*, 112. https://doi.org/10.1016/j.erss.2024.103513

Walker, G. *et al.* 2010. Trust and community: Exploring the meanings, contexts and dynamics of community renewable energy. *Energy policy*, 38(6): 2655–63. https://doi.org/10.1016/j.enpol.2009.05.055.

Wang, X. and Lo, K. (2021). Just Transition: A Conceptual Review. *Energy Research and Social Science, 82.* https://doi.org/10.1016/j.erss.2021.102291

Wang, N. et al. 2024. Synergies between the circular economy and carbon emission reduction. *Science of the Total Environment*,951. https://doi.org/10.1016/j.scitotenv.2024.175603.

Weber, M., 1978. *Economy and Society.* Berkeley: University of California Press

Weipert-Fenner, I. 2025. Green Hydrogen Production in Tunisia: The Interplay of Old and New Lines of Conflict. Peace Research Institute Frankfurt. Available at: https://www.prif.org/fileadmin/Daten/Publikationen/Prif_Reports/2025/prif_2_25_barrierefrei.pdf. (Accessed: 20 October 2025)

Wewerinke-Singh, M. 2021. A human rights approach to energy: Realizing the rights of billions within ecological limits. *Review of European, Comparative & International Environmental Law*, 31(1), 16–26. https://doi.org/10.1111/reel.12412.

Willand, N., & Horne, R. 2018. "They are grinding us into the ground"–The lived experience of (in) energy justice amongst low-income older households. *Applied Energy*, *226*, 61–70.

Williams, S. and Doyon, A. (2019). Justice in Energy Transitions. *Environmental Innovation and Societal Transitions,* 31. https://doi.org/10.1016/j.eist.2018.12.001;

Woehrle, L.M. 2022. Gender Studies. *Encyclopedia of Violence, Peace, & Conflict.* Elsevier.

WoMin. 2025. 'About Us.' Available at: https://womin.africa/about-us/ (Accessed: 15 June 2025);

WoMin 2025. 'Women Building Power.' Available at: https://womin.africa/category/blog/women-building-power/. (Accessed: 15 June 2025).

WoMin. 2025. 'Resisting Green Extractivism.' Available at: https://womin.africa/category/green-extractivism/. (Accessed: 27 June 2025)

Wood, N. (2023). Problematising Energy Justice: Towards Conceptual and Normative Alignment. *Energy Research & Social Science, 97.* https://doi.org/10.1016/j.erss.2023.102993

World Bank. 2021. Peru - Mining Sector Diagnostic (English). Available at: http://documents.worldbank.org/curated/en/448141632473923087 (Accessed: 8 August 2025).

World Bank Group. 2024.A virtuous cycle of green, affordable, and financially viable energy production. Available at: https://www.worldbank.org/en/programs/teri/brief/green-affordable-and-viable-energy-production-tunisia#:~:text=The%20Government%20of%20Tunisia%20(GoT,2030%2C%20against%203%25%20currently. (Accessed: 20 October 2025)

World Health Organization. Regional Office for Europe. 2016. World Health Organization. Regional Office for Europe. Available at: https://iris.who.int. (Accessed: 20 Aug 2025).

World Energy Council. National Hydrogen Strategies. 2021. Available at: https://www.worldenergy.org/assets/downloads/Working_Paper_-_National_Hydrogen_Strategies_-_September_2021.pdf. (Accessed: 11 August 2025).

World Nuclear Association. 2024. Advanced Nuclear Power Reactors. London: World Nuclear Association. Available at: https://world-nuclear.org/information-library/current-and-future-generation/outline-history-of-nuclear-energy. (Accessed: 11 August 2025).

World Nuclear Association. 2025. Landmark Ministerial Declaration puts nuclear energy at the heart of action on climate change. Available at: https://world-nuclear.org/net-zero-nuclear/news/landmark-ministerial-declaration-puts-nuclear-energy-at-the-heart-of-action-on-climate-change. (Accessed: 6 November 2025)

World Trade Statistics. 2025. Available at: https://www.wto.org/english/res_e/statis_e/world_trade_statistics_e.htm. (Accessed: 20 August 2025)

Zengin, O.Z. 2024. The Energy Justice Imperative for Clean Energy Storage Alternatives In: Heffron, R.J. and de Fontenelle, L. (Eds). The Power of Energy Justice & the Social Contract. Just Transitions. Palgrave Macmillan, Cham. 222

Zhang, C. 2025. EU to discuss potential changes to RFNBO rules as part of 2026 revision of bloc's hydrogen strategy. Hydrogen Insight. Available at: https://www.hydrogeninsight.com/policy/eu-to-discuss-potential-changes-to-rfnbo-rules-as-part-of-2026-revision-of-blocs-hydrogen-strategy/2-1-1882918. (Accessed: 20 October 2025

Index

A. Phillips, L. Kaschny (eds.), *Global Perspectives on the Circular Economy in the Just Transition*, Just Transitions,
https://doi.org/10.1007/978-3-032-18342-2

F

G

H

I

S

T

GPSR Compliance

The European Union's (EU) General Product Safety Regulation (GPSR) is a set of rules that requires consumer products to be safe and our obligations to ensure this.

If you have any concerns about our products, you can contact us on ProductSafety@springernature.com

In case Publisher is established outside the EU, the EU authorized representative is:

Springer Nature Customer Service Center GmbH
Europaplatz 3
69115 Heidelberg, Germany

Batch number: 10732017

Printed by Printforce, the Netherlands